ALL THE
RIGHT
ANGLES

A FIREFLY BOOK

Published by Firefly Books Ltd. 2013

First printing

Publisher Cataloging-in-Publication Data (U.S.)

Levy, Joel.
All the right angles : from gear ratios to calculating odds : mathematics in the world of sport / Joel Levy ; illustrated by Peter Dawson.
[224] p. : col. ill., photos. ; cm.
Includes bibliographical references and index.
ISBN-13: 978-1-77085-196-2
1. Sports – Mathematics. I. Dawson, Peter. II. Title.
796.0151 dc23   GV706.8.L489 2013

Library and Archives Canada Cataloguing in Publication

Levy, Joel
        All the right angles : from gear ratios to calculating odds : mathematics in the world of sports / Joel Levy ; illustrated by Peter Dawson.
Includes bibliographical references and index.
ISBN 978-1-77085-196-2
        1. Sports--Mathematics. I. Title.
GV706.8.L49 2013   796.01'51   C2013-900927-2

Published in the United States by
Firefly Books (U.S.) Inc.
P.O. Box 1338, Ellicott Station
Buffalo, New York 14205

Published in Canada by
Firefly Books Ltd.
50 Staples Avenue, Unit 1
Richmond Hill, Ontario L4B 0A7

Cover design by Erin R. Holmes/Soplari Design
Interior design by Grade Design

Printed in China by 1010 Printing International Limited

Conceived and produced by
Quid Publishing
Level 4, Sheridan House
114 Western Road
Hove BN3 1DD
England
www.quidpublishing.com

ALL THE RIGHT ANGLES

From Gear Ratios to Calculating Odds: Mathematics in the World of Sports

Joel Levy

Illustrated by Peter Dawson

FIREFLY BOOKS

Contents

Introduction

Did you know that the oldest known soccer ball dates to 1540 C.E., that the British cyclist Bradley Wiggins climbed the equivalent of a 62-mile-high mountain in training for the 2012 Tour de France, or that a pronghorn antelope would finish galloping the 800-meter race while the world record-holding human was only one-third of the way round the running track?

Did you know that a baseball batter has just over one-fifth of a second to decide whether, how hard and at what angle to hit the ball, and that getting the timing wrong by 7 milliseconds will make the difference between a home run, a pop fly or a strike?

Did you know that an average man can hold his breath for over 20 min, that badminton is the fastest racket sport in the world and that, statistically speaking, Australian cricket player Donald Bradman's batting average is the single most impressive record in sports?

Is it all simply fascinating trivia, or do these facts and figures have any greater significance for the world of sports? Sports are all about motion, power, energy and momentum. The people and objects involved move in lines and curves, their progress determined by geometry. Any definition of sports begins with the term "physical," as in "governed by physics." To regulate and score sports, times are recorded, points counted, scores calculated. Every one of these phenomena, processes and events is governed by numbers, from the energy involved in propelling a ball to the calculation of a coefficient that determines tournament seeding. It is not possible to understand sports without numbers, and without understanding it is impossible to improve.

At elite levels, winning is determined by the tiniest of margins and precise analysis is essential to the incremental improvement of performance. Knowing the numbers can give you the edge. This is increasingly recognized in professional sports, where the introduction of statistical recording and analysis systems, such as

Sabermetrics in baseball and Prozone in soccer have made major contributions to athletic success and placed mathematics firmly in the heart of the locker room and the manager's office. Athletes should be interested in the mathematics of sports for this simple reason if nothing else: the application of science can help them win.

The traffic between sports and mathematics goes both ways, because sport offers a unique and engaging way into mathematics. Everyone, whether they realize it or not, is a sort of 'folk mathematician', unconsciously relating quantities and magnitudes, angles and vectors. But too many people switch off as soon as the maths in their life becomes explicit, and sport has an unrivalled power to turn them on to the power and beauty of numbers.

The fascination of numbers, geometry and statistics will hopefully shine through the pages of this book and you will never think the same way about hitting, or watching someone else hitting, a nine iron, swinging at a curveball, winding up for a slap shot or curling in a free kick.

Glossary

FORCE

A force is a push or pull on an object as the result of interaction with another object. When the racket hits the ball it exerts force, but when the ball leaves the racket there is no more force being exerted. To accelerate an object from rest to motion, or decelerate a moving object to a stop, requires force.

MOMENTUM

Momentum is mass in motion. It is a quantity that depends on mass and speed. Momentum can be a measure of how hard it is to stop something or how much energy it carries.

GRAVITY

A force of attraction that acts between any two objects (i.e., things with mass). Gravity is relatively weak, so only very big objects, like the Earth, exert noticeable gravity. The Earth's gravity is always pulling at objects, giving them weight.

WEIGHT

The downward force exerted by a mass because of gravity. Although weight depends on mass (so that the terms are often used interchangeably), it is not the same thing. Properly speaking weight should be expressed in newtons (force units) rather than pounds.

INERTIA

Tendency to resist acceleration, whether this means speeding up an object at rest or slowing down a moving one. The more mass something has, the more inertia it has.

TORQUE

Rotational force, or more specifically, a measure of how much a force acting on an object makes it rotate.

MASS

How much matter an objects contains. In space, mass is weightless, but on Earth gravity gives mass weight. Mass determines inertia.

POWER

The rate at which energy is transformed from one form into another, or at which work (movement of something by application of a force) is done.

POTENTIAL ENERGY

The energy something has because of its position, usually meaning because it has been raised up and now has room to fall, allowing the force of gravity to accelerate it, transforming potential energy into kinetic energy.

KINETIC ENERGY

Energy of motion. The faster something is moving, the more kinetic energy it has.

CONSERVATION

In physics terms, the principle that something (usually energy or momentum) cannot disappear or appear out of nowhere, it can only be transformed. For instance, in a squash ball and the wall of the court, the total amount of energy or momentum before and after impact must be the same.

Equations

MOMENTUM = MASS x VELOCITY

Since momentum is "mass in motion," momentum is quantified by multiplying mass by velocity. Velocity is not the same as speed—it is a vector, meaning it has direction, as does momentum.

FORCE = MASS x ACCELERATION

Newton's laws of motion explain that force is what accelerates a body at rest into motion, or decelerates a moving body until it comes to rest. The unit of force is the newton (N).

WEIGHT = MASS x ACCELERATION DUE TO GRAVITY (G)

Weight is a type of force exerted by bodies in a gravity field. When mass and G are expressed in "standard units" (i.e., the metric system), G is roughly 10. So to approximate someone's weight in newtons from their mass in kilograms, just multiply by ten (1 kg = 2.2 lb).

POTENTIAL ENERGY = MASS x HEIGHT x GRAVITY

The formula for potential energy shows how raising a mass to a certain height gives it stored or potential energy, measured in joules.

Power and Motion

ALL SPORTS ARE MOTION, ACTIVITY, THE EXPENDITURE OF ENERGY AND ITS TRANSFORMATION INTO OTHER FORMS. ALL OF THESE PROCESSES AND PHENOMENA ARE GOVERNED BY FUNDAMENTAL LAWS OF NATURE, FROM NEWTON'S LAWS OF MOTION TO THE LAWS OF THERMODYNAMICS. SPORTS CAN HELP TO ILLUSTRATE AND ILLUMINATE THESE PRINCIPLES, AND AT THE SAME TIME THESE PRINCIPLES CAN IMPROVE BOTH UNDERSTANDING AND PERFORMANCE OF SPORTS.

Bouncing Back to You: Basketball and Conservation of Energy

If you take a basketball—or any ball, from a squash ball to a soccer ball—and drop it on the ground, it will bounce back up. This is because of a very important principle called the conservation of energy, which says that energy cannot disappear.

When you pick up a ball and raise it off the ground, you are working against gravity and giving the ball potential energy. Potential energy (PE) is a measure of how much energy an object has thanks to mass (M), height (H) and gravity (G): PE = M x H x G.

It is easy to work out the PE (measured in joules, J) of a basketball using this simple equation. Basketballs weigh around 0.65 kg (23 oz), while the average height of a National Basketball Association (NBA) player is around 2 m (6 ft 7 in). Gravity accelerates mass at 9.8 m/s^2 (32 ft/sec^2), so if we imagine an NBA player dropping a basketball from about shoulder height, the PE of the ball is 0.65 x 1.8 x 9.8 = 11.47 J.

As the ball falls to the ground, the potential energy is converted into kinetic energy, and these joules of kinetic energy cannot simply disappear when the ball hits the ground. Thanks to the elasticity of the ball, most of them are converted into kinetic energy, propelling the ball back up. According to international specifications, a regulation basketball must bounce at least 1.3 m (51.2 in) when dropped from a height of 1.8 m (70.9 in) on a hard surface. We can calculate the PE the ball will have when it reaches the top of its 1.3 m bounce: 0.65 x 1.3 x 9.8 = 8.28 J.

If energy cannot disappear, what happened to the (11.47 - 8.28 =) 3.19 J difference between the before and after bounce? It has been lost due to air resistance, friction etc., which dissipate energy as heat, sound etc. The basketball itself will be a minute bit hotter, and although this sort of change will be imperceptible, you can easily feel the difference in a squash ball that has been knocked around for a while (see page 78).

7 ft 7 in

Height of Manute Bol, tallest NBA player ever

0.5 in

Maximum deformation of a properly inflated basketball when it hits the ground

0.016 s

Roughly the amount of time a bouncing basketball is in contact with the ground

↓ 650 N

Maximum force exerted by a bouncing basketball on impact

THE RELATIVE BOUNCE OF BALLS

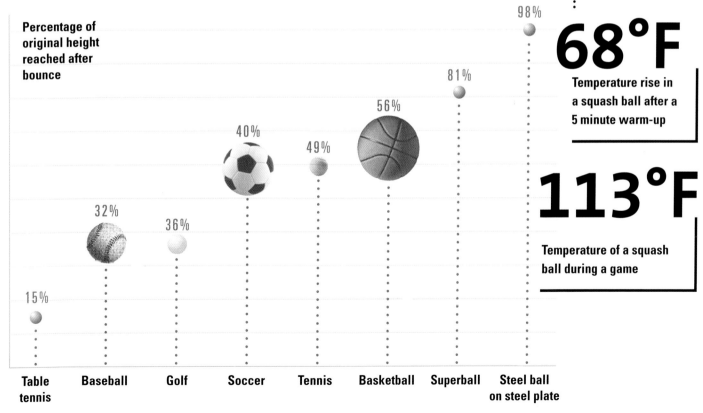

Percentage of original height reached after bounce

							98%
						81%	
				56%			
			40%	49%			
	32%	36%					
15%							
Table tennis	Baseball	Golf	Soccer	Tennis	Basketball	Superball	Steel ball on steel plate

68°F

Temperature rise in a squash ball after a 5 minute warm-up

113°F

Temperature of a squash ball during a game

Turn Up the Heat: Baseball and Momentum

80 N/m²

Torque acting on the elbow of a pitcher throwing at 100 mph

3.4 mi

Average distance traveled by baseball pitches during a Major League Baseball (MLB) game in 2009

Watch a baseball pitcher winding up for a fastball, and you will see that he makes some strange shapes with his body: lifting one leg before planting it down on the mound, twisting his torso and swinging his hips, cocking his arm and throwing out his elbow, and whipping his forearm through and his wrist up and over. All of these contortions are intended with one aim in mind— to transfer momentum to the ball.

What is momentum? We will meet this concept repeatedly throughout the book. Momentum is mass (M) multiplied by velocity (v), or, in more familiar terminology, weight x speed. A small object traveling slowly has a small amount of momentum, while a big object traveling fast has a great deal of momentum. Momentum can be transferred from one object to another, and the crucial point, as far as the science of sports is concerned, is that momentum is conserved during this transfer. So when a large object transfers its momentum to a small object, the small object must end up with the same total momentum. Since the mass of the small object does not increase (except in terms of special relativity—see page 16), its speed must go up instead. This is the principle behind almost all ball sports, in which a ball is accelerated to high speeds through the application of momentum.

In a baseball throw the pitcher generates momentum by putting the entire mass of his body in motion. His body may not move that fast, but compared to a baseball his mass is far greater. The pitcher's complicated movements transfer that whole body momentum into progressively smaller body parts, from his whole body (when the legs move), to the torso, to the arm, to the hand. By the time his hand comes through to pitch the ball, his body has acted like a whip, concentrating momentum into a small zone. This allows him to release the ball at a high speed, with most of the momentum generated by his body weight transferred to the ball.

105 mph

Fastest fastball ever pitched in MLB, by Aroldis Chapman of the Cincinnati Reds, in September 2010

74/159

Proportion of 100+ mph pitches thrown by Chapman in the 2010 MLB season to September

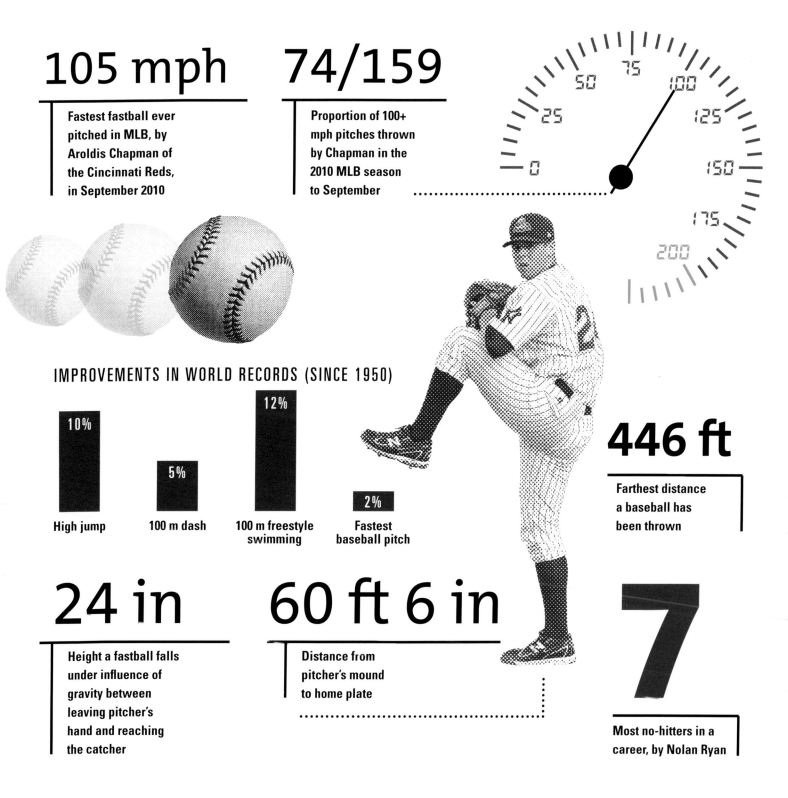

IMPROVEMENTS IN WORLD RECORDS (SINCE 1950)

- **10%** High jump
- **5%** 100 m dash
- **12%** 100 m freestyle swimming
- **2%** Fastest baseball pitch

446 ft

Farthest distance a baseball has been thrown

24 in

Height a fastball falls under influence of gravity between leaving pitcher's hand and reaching the catcher

60 ft 6 in

Distance from pitcher's mound to home plate

7

Most no-hitters in a career, by Nolan Ryan

Relatively Heavy: Baseball and Relativity

30

Mass and energy are directly related, so when a ball gains speed (and therefore kinetic energy) it also gains mass. The faster the ball goes, the heavier the ball gets. How is this possible? The key is Albert Einstein's famous equation, $E=mc^2$. This equation shows how energy (E) relates to mass (m), with the relationship governed by the constant (c), which is the speed of light in a vacuum.

The speed of light is colossal, and when squared it gets even bigger, so the equation tells us that a small amount of mass is equivalent to an enormous amount of energy. This is why a tiny amount of uranium in an atom bomb can produce such a massive explosion. When the equation is restated—$m=E/c^2$—it shows that energy is equivalent to a vanishingly tiny amount of mass. In practical terms this means that although a baseball does indeed get heavier when it goes faster, the increase in mass is so minuscule it makes no difference to the pitcher or batter. To put it another way, the physics of baseball is governed by Issac Newton, not Einstein.

Increase in mass with increasing velocity is why nothing can go faster than light. The closer to light speed an object gets, the more massive it becomes. The more massive something becomes, the more energy is needed to accelerate it. But the more energy it has, the more massive it becomes. This vicious circle means that an object such as a baseball can never reach light speed, which must come as a relief to hapless batters who have enough trouble picking up a hit against a 90 mph fastball.

Number of H-bombs equivalent to your mass energy

$$E=mc^2$$

0.00000000000007 oz

Amount by which a baseball thrown at 100 mph increases in mass

0.000000000004 oz

Difference in mass between a charged and a discharged battery due to the mass of the energy involved

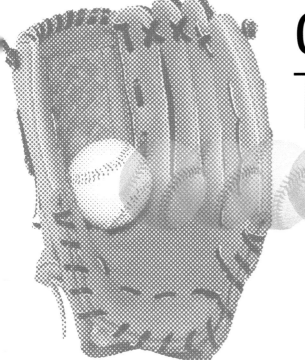

0.5 s

Time for a 90 mph fastball to reach home plate

0.4 s

Time for a 95 mph fastball to reach home plate

0.2 s

Normal human reaction time

1 mph

Rate at which a ball slows down after it leaves the pitcher's hand, per 7 ft

8 mph

Amount by which a ball has slowed by the time it crosses the plate

Throwing a Curve: Ball Games and the Magnus Effect

Making a ball curve through the air is a vital skill in many ball games, from baseball and cricket to tennis, golf and soccer. The exact physics behind it are complex and differ from sport to sport, but the basic phenomenon is called the Magnus effect.

The Magnus effect is the process by which a spinning ball deflects airflow to create a force pushing the ball. It is easiest to explain if you imagine a ball that is spinning but staying in one spot, with air flowing over it. On one side of the ball the air will be flowing in the same direction as the spin, while on the other side the spin will be in the opposite direction of the airflow. Where the two line up the spinning ball will accelerate the air, and where they oppose, the spinning ball will slow down the air. This makes the air flow faster on one side than the other, so that it is deflected toward the slow side, in the same way that if you are rowing and your left oar is rowing faster than your right one, the boat will turn right.

According to Newton's laws of force, every action has an equal and opposite reaction, so if the air is deflected to the right, there will be a force acting on the ball to push it to the left. This makes the spinning ball swerve or curve in flight.

66%

Chance of a batter missing a curveball thrown by Adam Wainwright, one of the best curveball pitchers in MLB, when on a 2–2 count, 2009 season

600 rpm

Spinning speed of
a typical curveball

20 ft

Distance from home plate at which
a curveball switches from the batter's
central to peripheral vision (accounting for
the illusion of "breaking," when a curveball
appears to drop at last moment)

5

Approximate number
of times a curveball
spins between pitcher
and home plate

14 in

Apparent break
(distance by which ball
appears to deviates
from a straight line) of
a typical curveball

3.5 in

Actual break of
same curveball

19 in

Apparent break
achievable by
pitching curveball
with four seams
meeting the wind

7.5 in

Apparent break of
a curveball pitched
with two seams
meeting the wind

The Hardest Task in Sports: Timeline of a Baseball Swing

38%

.250

The job of the batter in baseball is sometimes called the hardest task in sports. This is because the batter has to hit a ball with a narrow stick, and the ball is traveling at high speed and moving sideways through the air.

We've already seen how pitches in MLB can be extraordinarily fast and curvy, but just to recap: it takes c. 400 milliseconds (0.4 seconds) for the ball to travel from the pitcher to the home plate. This short stretch of time breaks down as follows for the batter:

0–100 milliseconds: nerve signals travel from the eye to the brain, enabling the batter to see where the ball is going

100–175: the brain processes the information to decode speed and direction

175–200: the batter has just 25 milliseconds to decide whether to swing or not

200–225: he has another 25 milliseconds to decide what kind of swing—e.g., high or low, inside or outside

225–250: nerve signals travel from brain to legs to start stride; this takes at least 15 milliseconds but if it takes more than 20 the batter will have missed his opportunity, because it takes 150 milliseconds to complete a swing

250–300: the batter has started his swing, but he can still pull out if he changes his mind

c. 300: if the pitch is a curveball it breaks sharply around now, about 15 ft from home plate

300–400: the bat is moving too fast for the batter to stop it

400: if the batter has timed and directed his swing correctly, he will meet the 3 in wide spinning ball within ¾ in of dead center of his bat.

No wonder they say batting is a tough task. Do the statistics agree? In fact, even the best batters in MLB have a "success rate," as defined by their batting average (see page 128), of just one in four, making them among the worst-performing elite athletes in the world, despite being the highest paid.

80 mph

Speed of bat on impact with ball

0.05 s

Window for deciding when to swing

9 hp

Energy delivered to ball

0.007 s

Margin of error for batters. Swing 7 ms (0.007 s) too soon and the ball will go foul down the left field side; 7 ms too late and ball will go foul to the right (for right-hand batters)

0.001 s

Duration of contact between bat and baseball

Baseball

88.5 × 125 m

Outfield

DIMENSIONS OF FIELD

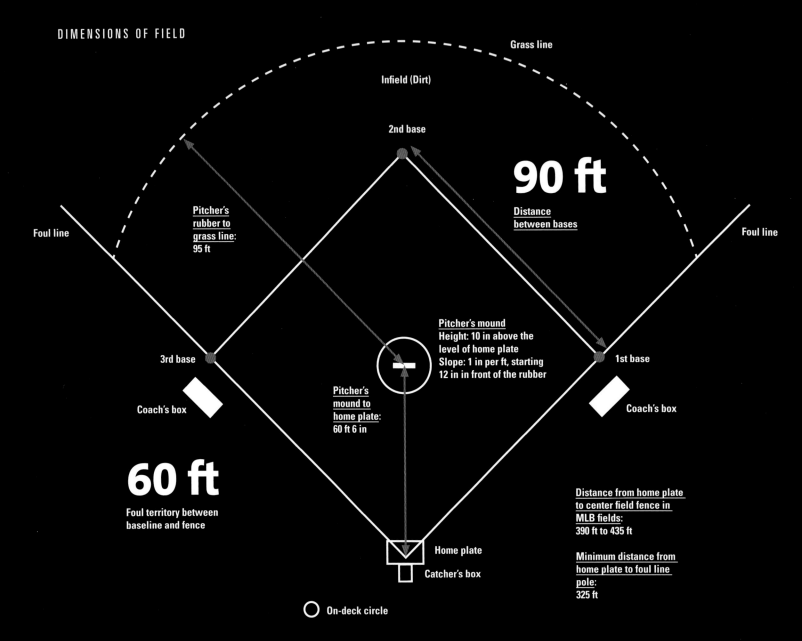

Grass line

Infield (Dirt)

2nd base

90 ft

Distance
between bases

Foul line

Foul line

Pitcher's
rubber to
grass line:
95 ft

Pitcher's mound
Height: 10 in above the
level of home plate
Slope: 1 in per ft, starting
12 in in front of the rubber

3rd base

Pitcher's
mound to
home plate:
60 ft 6 in

1st base

Coach's box

Coach's box

60 ft

Foul territory between
baseline and fence

Distance from home plate
to center field fence in
MLB fields:
390 ft to 435 ft

Minimum distance from
home plate to foul line
pole:
325 ft

Home plate

Catcher's box

On-deck circle

BASEBALL

Diameter: 3 in

Circumference: 9 in

Weight: 5 ¼ oz

3 in

BEST BATTING AVERAGES IN MLB HISTORY

Ty Cobb .366
Rogers Hornsby .359
Joe Jackson .356
Ed Delahanty .346
Tris Speaker .345
Ted Williams .344
Billy Hamilton .344
Dan Brouthers .342
Babe Ruth .342
Harry Heilmann .342

BAT
(solid wood, usually ash)

**Diameter at barrel
(thickest part):**
2 ¾ in

Length: not more
than 42 in

2 ¾ in

42 in

High Impact: Football and Newtonian Physics of Collision

When Newton formulated his laws of motion he spoke of bodies at rest and in motion. He probably didn't have actual human bodies in mind, but his laws of force and motion apply to us as well as everything else in the macroscopic universe. This is nowhere better illustrated than in football, in which direct unchecked full-body contact collisions are not simply allowed but positively encouraged. Looking at the mathematics of a celebrated example of such contact brings home the way in which football players literally put their bodies on the line.

In October 2010, the Atlanta Falcons played the Philadelphia Eagles in a National Football League (NFL) game marked by a stunning clash between two players colliding head-on. DeSean Jackson received an angled pass and was promptly flattened by Dunta Robinson. Both players were knocked back and into the air, with Jackson suffering concussion. According to an analysis by Jessie Wynen, this collision was notable because the two players involved were of similar build and mass, and the impact was head-on with very little sideways deflection, making it possible to calculate the enormous forces involved.

Force = mass (M) x acceleration, so to calculate the impact force that acted on Jackson we need to know his mass and acceleration. His stats at the time list his weight as 175 lb (79.56 kg). His acceleration can be deduced from the fact that his speed was reduced to zero in the space of time the impact occurred. His speed, in turn, can be worked out from his 40-yd dash stats, which show that he could run 40 yd (37 m) in 4.29 seconds, giving him a speed of 30 fps (8.53 m/s).

Change in velocity (kg) / time (s) = Acceleration (m/s^2) $$\frac{8.53}{0.2} = 42.63 \text{ m/s}^2$$

Mass (kg) x acceleration (m/s^2) = Force (N) $79.56 \times 42.63 = 3391.64 \text{kg m/s}^2 \sim 3400 \text{ N}$

30 fps
DeSean Jackson's speed before impact

0.2 s
Duration of tackle impact

175 lb
Jackson's mass

760 lb
Mass equivalent of 3400 N, the force of impact of Dunta Robinson on DeSean Jackson

3,000 lb
Maximum impact body can sustain if protected by player's body armor

G-LOADS

Walking: 1.0

Sneezing: 2.9

Shuttle launch: 3.0

Roller coaster: 5.0

Fighter jet roll: 9.0

Concussion: 100

150 g
g-force of an extreme impact in football

100 g
Acceleration threshold for concussion

100+
Number of concussions recorded each season in the NFL

The Perfect Pass: Football and Aerodynamics

Like a thrown basketball (see page 12), a football follows a parabolic path when the quarterback throws a pass. The quarterback throws the ball at an angle of just under 45° to get the maximum distance, but the interaction of the forward velocity, the upward launch trajectory and the downwards acceleration of gravity combine to produce a parabolic path.

A quarterback is not just looking to throw the ball as hard as he can. The shape of a football means that simply hurling it into the air will achieve neither the maximum distance possible, nor the accuracy needed to pick out receivers. A football is a prolate spheroid, not a sphere. If simply thrown upwards it will be at an angle with its nose pointing in the air, and as it comes down its nose will still be in the air. Air resistance will make the ball tumble and this is in turn will decrease the accuracy and length of the throw. To drill long, accurate passes downfield, the quarterback uses the laces to spin the ball, allowing him to throw a spiral pass.

In a perfectly thrown pass the football spins around its axis of symmetry (imagine a pole stuck through the length of the ball). As a result, the football tilts in flight so that its axis of symmetry is always pointing in the direction the ball is going (in technical terms, the axis is tangent to the trajectory). Spin turns the ball into a gyroscope. As soon as the axis of the football starts to differ from the direction of flight, aerodynamic drag acting against the line of flight creates torque (force acting to rotate something), and this moves the ball back into line until the torque stops acting. This happens all along the path of the ball, so that it is constantly kept in line with its direction of flight. What's more, the aerodynamic qualities of the spinning ball act a bit like a wing, generating lift and helping the ball stay aloft longer and travel farther.

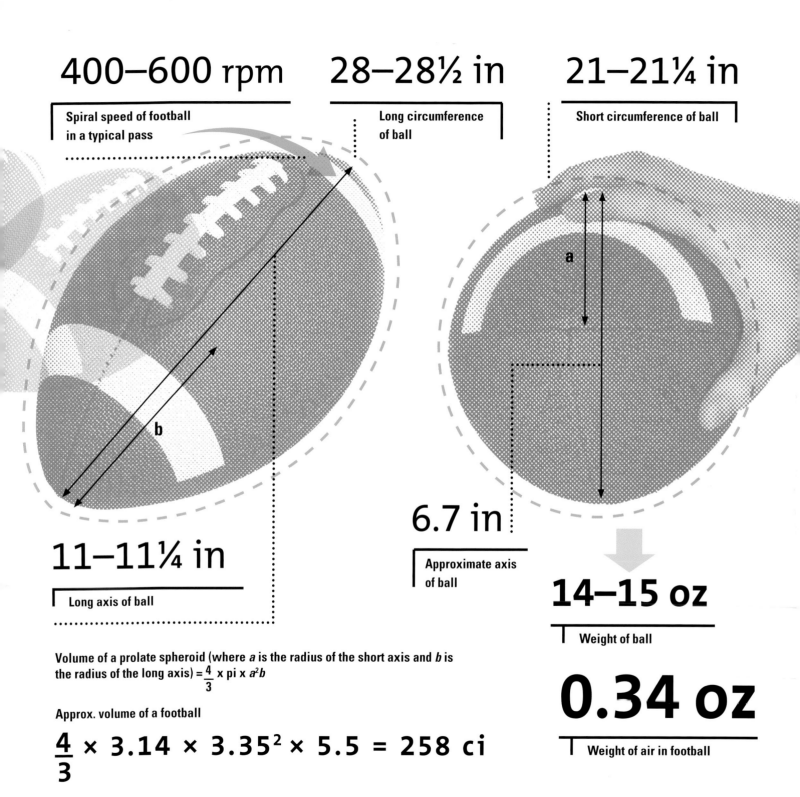

400–600 rpm

Spiral speed of football
in a typical pass

28–28½ in

Long circumference
of ball

21–21¼ in

Short circumference of ball

b

a

11–11¼ in

Long axis of ball

6.7 in

Approximate axis
of ball

14–15 oz

Weight of ball

Volume of a prolate spheroid (where a is the radius of the short axis and b is
the radius of the long axis) = $\frac{4}{3}$ x pi x $a^2 b$

Approx. volume of a football

$$\frac{4}{3} \times 3.14 \times 3.35^2 \times 5.5 = 258 \text{ ci}$$

0.34 oz

Weight of air in football

Football

Players on the field (per team): 11

Length of game: 60 min, played in four 15 min quarters

Length of halftime: 12 min

Number of time-outs per half per team: 3

DIMENSIONS OF FIELD

100 yd

Length of field: 100 yd

10 ft

Height of crossbar of goal post: 10 ft

30 ft

Length of each end zone: 30 ft

160 ft

Width of field: 160 ft

1,948

Number of active NFL players in 2011

481,156 lb

Total weight of NFL players in 2011 = approx 1.2 adult male blue whales

NUMBER OF NFL PLAYERS OVER 300 LB

1970	1
1980	3
1990	94
2000	301
2010	532

260 lb

Weight of Green Bay Packers' largest player in 1967

13

Number of 300 lb + players in Green Bay Packers in 2010

247 lb

Average weight of NFL player in 2011

Spares and Strikes: Bowling and Algebra

16 lb

The weight of a bowling ball is not evenly distributed. As it rolls along gravity applies greater force on one side than the other, and this causes it to deviate from a straight line. If you were looking down at the ball from above and you traced its path along the lane, you would produce not a straight line but a curve. This raises a problem for the bowler: if the ball will curve to the left by a certain number of degrees for every foot that it travels, how far to the right should you aim to hit a bowling pin at the end of the lane?

Answering this question involves the mathematics of curves. A line drawn on a graph can be described in terms of its position relative to the x (horizontal) and y (vertical) axes. This gives us algebra. For instance, if you were looking down at someone bowling a ball that started 16 in to the left of the center of the foul line (which we call our x-axis), and it went in a straight line parallel to the gutter, the line the ball traced could be easily described as x=16. If the ball was thrown in a straight line but at an angle so that it moved 8 in to the left for every 16 in it traveled along the lane, the line it traced could be described as $y = \frac{1}{2} x$

But a ball that traces a curved path gives a different challenge. To describe a curve, it is necessary to introduce an exponent. For instance, the line described by $y=x^2$ is a curve where the line goes up by the square of the amount it goes across. In the case of the curving bowling ball, the curve might be something like the diagram on the next page.

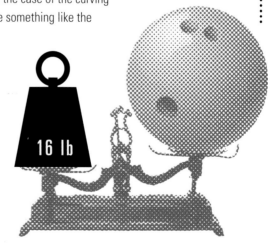

16 lb

$$\frac{\text{Distance (ft)}}{\text{speed (fps)}} = \text{time (s)} \quad \frac{60}{28} = 2.14$$

Mass (kg) x speed (m/s) = momentum (kg m/s)

$$7.25 \times 7.6 = 55.1$$

2.14 s

Time taken from release to impact for bowling ball to travel along lane

60 ft

Length of lane

25 fps

25 fps (7.6 m/s) = 17 mph: speed of ball on impact with pins

28 ft

Average approximate speed of bowling ball

31 fps

Speed of bowling ball at release

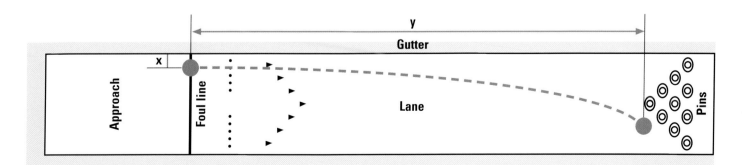

Weight Watchers: Tour de France and Human Power Output

The Tour de France is seen by many as the ultimate test in sports. The 2012 edition, for example, covered 2,173 miles in just 23 days, and included nine mountain stages but just two rest days. The amazing demands on the bodies of the riders means that Tour cyclists push the boundaries of human physiology. Especially interesting are measurements of the riders' VO_2 Max and their power output. These metrics show how much oxygen they can take in and how much power they generate per pound of weight.

Bradley Wiggins, winner of the 2012 Tour, achieved this milestone by changing his body, specifically his weight, so that he could marginally increase his power output per pound. During the 2007 Tour, Wiggins weighed 172 lb. By 2009, when he finished fourth, he had shed 15 lb, and for 2012 he somehow shaved off another 3 lb, competing at a weight of just 152 lb.

Why go through such a punishing regime of weight loss? Grand tour winners need to compete over shorter distances on flat ground in the time trials, but, crucially, they also need to be able to keep up with the pack over the mountains. To manage this feat they need a power output per kilogram of around 7 W. Wiggins was already close to his power output limit, although he did manage to boost it from an average of around 460 W during the 2011 World Time Trial Championships to around 475 W in the 2012 Tour, peaking at an astonishing 492 W during the prologue stage. The other way to change the arithmetic and boost output per pound is to lose weight, hence the challenging weight loss program.

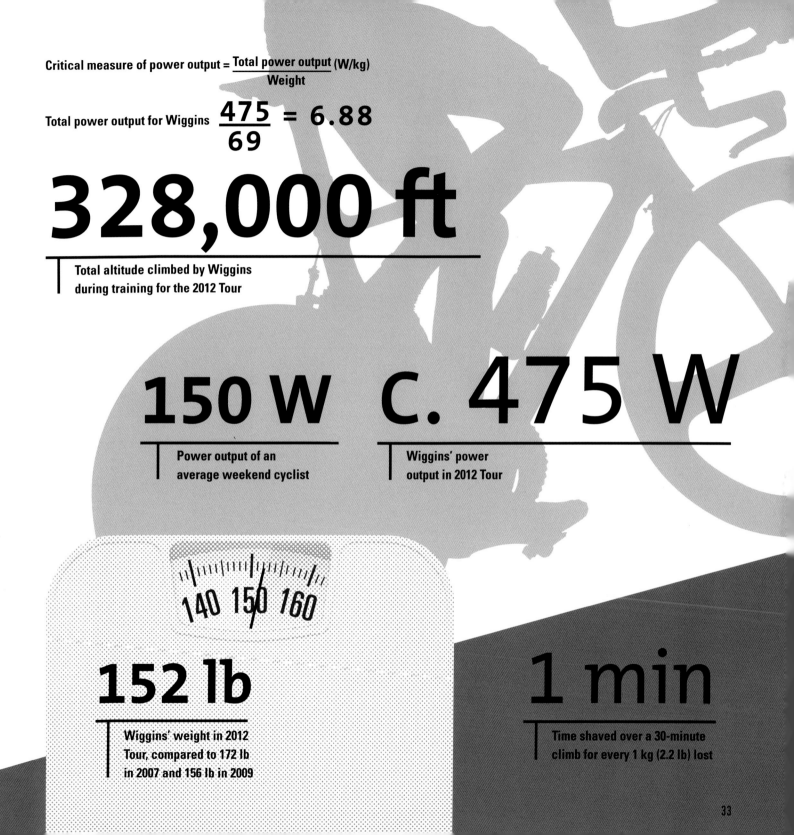

Critical measure of power output = $\dfrac{\text{Total power output}}{\text{Weight}}$ (W/kg)

Total power output for Wiggins $\dfrac{475}{69}$ = 6.88

328,000 ft

Total altitude climbed by Wiggins
during training for the 2012 Tour

150 W c. 475 W

Power output of an
average weekend cyclist

Wiggins' power
output in 2012 Tour

140 150 160

152 lb

Wiggins' weight in 2012
Tour, compared to 172 lb
in 2007 and 156 lb in 2009

1 min

Time shaved over a 30-minute
climb for every 1 kg (2.2 lb) lost

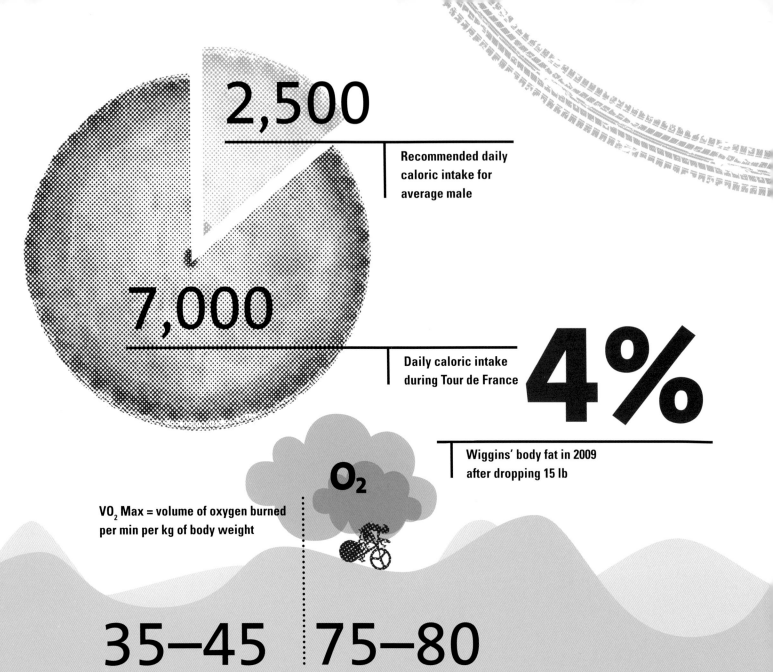

2,500

Recommended daily caloric intake for average male

7,000

Daily caloric intake during Tour de France

4%

Wiggins' body fat in 2009 after dropping 15 lb

O₂

VO_2 Max = volume of oxygen burned per min per kg of body weight

35–45

Typical VO_2 Max for average 30-yr-old man

75–80

VO_2 Max for Wiggins

WIGGINS' 2012 TOUR

2,173 mi
Total distance cycled

6,890 ft
Total distance climbed

31 mph
Average time-trialling speed

87 hr 34 min 47 s
Total time in saddle

Cheating: Cycling and Statistics

Drug cheats and anti-doping authorities are engaged in a constant battle that mimics the process of predator-prey evolution in nature. In nature, predators and their prey are locked into an evolutionary "arms race," in which increments in the predator's evolutionary fitness (i.e., the ability to catch prey) act as a selection pressure to drive improvements in the prey's evolutionary fitness (i.e., the ability to avoid being caught), and vice versa. If the next generation of cheetahs is faster, then the generation of pronghorn antelopes after that will be faster, and so on.

Similarly, drug cheats in sports are always looking for new products or new ways of using them, ways that the anti-doping authorities (ADAs) cannot yet detect. In response the ADAs develop new tests, driving the drug cheats to develop new ways of cheating, and so on ad infinitum.

But there is another way to spot potential cheats, and it depends on the mathematics of sports. At the highest echelons of sports, winning margins are tiny and the performances of elite athletes in terms of their stats (power output, lap times, power:weight ratio, etc.) should be closely aligned. If analysis of the stats shows some athletes performing out of this "normal" or "expected" range, they automatically fall under suspicion. This does not constitute proof, but it can direct investigations.

Cycling, particularly the Tour de France, is the sport under the most intense spotlight in relation to drug cheats, but other sports such as swimming have also made headlines in this arena, for example with the controversy over Chinese swimmer Ye Shiwen. Performance stats offer powerful evidence that the Tour de France, for instance, is now cleaner than it was in the 1990s and early 2000s—the "dirty years."

0.4 s

Enhancement in sprint capacity over 100 m dash for athletes injected with human growth hormone and testosterone in a 2010 study at St. Vincent's Hospital in Sydney, reported in the *Annals of Internal Medicine*

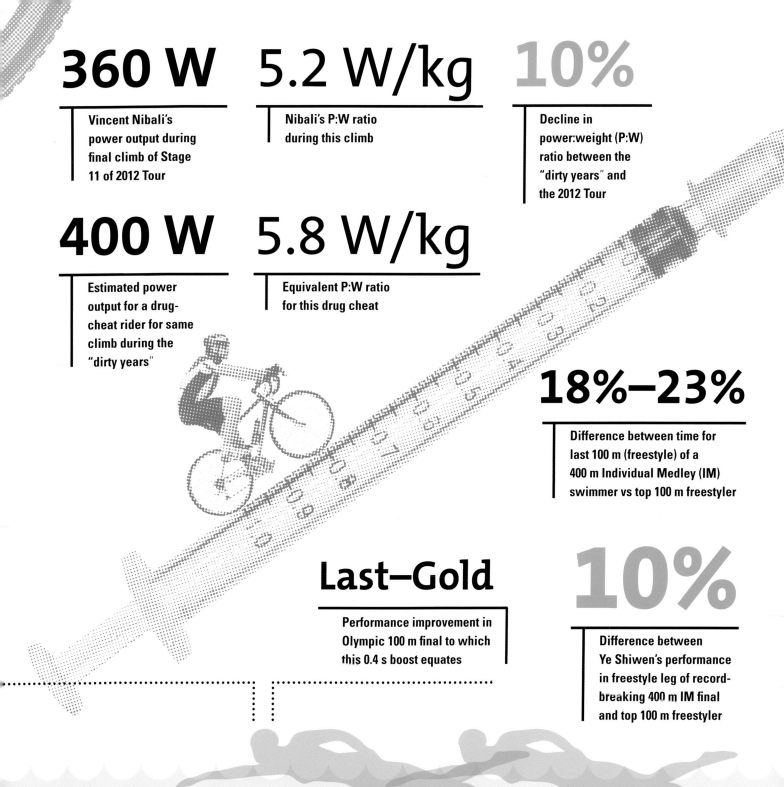

360 W

Vincent Nibali's power output during final climb of Stage 11 of 2012 Tour

5.2 W/kg

Nibali's P:W ratio during this climb

10%

Decline in power:weight (P:W) ratio between the "dirty years" and the 2012 Tour

400 W

Estimated power output for a drug-cheat rider for same climb during the "dirty years"

5.8 W/kg

Equivalent P:W ratio for this drug cheat

18%–23%

Difference between time for last 100 m (freestyle) of a 400 m Individual Medley (IM) swimmer vs top 100 m freestyler

Last–Gold

Performance improvement in Olympic 100 m final to which this 0.4 s boost equates

10%

Difference between Ye Shiwen's performance in freestyle leg of record-breaking 400 m IM final and top 100 m freestyler

Takedown: Judo and Levers

470 N

48 x 9.8 = 470 N: force needed to lift extra lightweight judoka

4,106 lb

(48 kg) Lightest class (extra lightweight) in women's judo

Force is what overcomes inertia and gets a mass moving. Gravity exerts a force on mass, accelerating it toward the center of attraction, and this is what gives us weight. Your mass might be 80 kg (176 lb), but this is not really a measure of your weight. It is more accurate to describe your weight in units of force, known as newtons (N). One newton is the force needed to give a mass of one kilogram (2.2 lb) an acceleration of 1 m/s^2 (3.28 ft/s^2). Force in N = mass in kg x acceleration in m/s^2, so on Earth, where acceleration due to gravity is 9.8 m/s^2 (32 ft/s^2), an 80 kg person has a weight of 80 x 9.8 = 784 N.

To lift the hypothetical 80 kg person you would need to overcome the force of gravity by applying 784 N of force in the opposite direction. Lifting a 100 kg (220 lb) person would take 100 x 9.8 = 980 N. In a judo fight between an 80 kg judoka (judo fighter) and a 100 kg judoka, the heavier man would seem to have the advantage. Not only would he probably be stronger (more muscle mass), but it would take less force for him to lift his lighter opponent.

But judo is a martial art based on the power of levers. Levers introduce an additional concept: torque or moment of force. Torque is a measure of rotational power, or the force that acts around a pivot—in other words the turning power of an applied force. Torque is determined by the length of the lever (the perpendicular distance from its line of action to the pivot) multiplied by the applied force. With a lever you could lift the 100 kg man using only 98 N of force if you applied the force to a point 10 m (32.8 ft) from the pivot: 98 N x 10 m = 980 N m (the measure of torque).

In a judo throw a small judoka can easily lift a big one off his feet by using elements such as hips and ankles as pivots, utilizing the power of levers.

2,000

Estimated number of fights by legendary Japanese judoka Mitsuyo Maeda (1880–1941)

2: losses by Maeda

0: losses by Maeda in fights where contestants wore judo uniforms

481 lb

Heaviest judoka at 2012 Olympics (and heaviest athlete in modern Olympic history), Ricardo Blas Jr of Guam

2,136 N

218 x 9.8 = 2,136 N: force needed to lift Blas

15 ft

Length of lever needed for lightweight judoka to balance Blas

388 lb

Weight of entire Japanese women's gymnastic team

5,760

Throws in 24 hours of nonstop throwing, aimed for by Turkish judoka Namik Ekin in world record attempt in July 2012

93 lb

Difference between weight of Blas and weight of entire Japanese women's gymnastic team

The Harder They Come: Rugby and Levers

Rugby is a contact sport in which ball carriers are stopped by tackling them. Since these ball carriers are often forwards, who tend to be very big and heavy, and since the players tackling them are often backs, who tend to be smaller and lighter, one might expect more tries to be scored by big players bulldozing their way across the line. But rugby players are taught the fine art of the tackle as one of the first elements of the game, and with the right technique the smallest player can stop the biggest with ease.

The key to tackling in rugby is the principle of the lever. Stopping a large ball carrier moving at speed by tackling him head-on requires the tackler to have as much momentum as the ball carrier, and ensures an almighty collision (see High Impact, page 24). For instance, in the modern game it is not uncommon to find 250 lb players who can run at speeds approaching 32 fps over short distances. In his playing days, for example, Jonah Lomu stood 1.96 m (6 ft 5 in) tall, weighed 119 kg (262 lb) and could run 100 m (328 ft) in 10.8 seconds. At this speed his momentum would be 119 kg x 9.26 m/s = 1,102 kg.m/s (7,957 lb.ft/sec).

But tackling the ball carrier head-on is unnecessary. By tackling him below his center of mass and clasping his feet together, you can immediately turn him into a lever that is pivoting around the ankles, whereupon his lateral momentum becomes angular momentum slamming into the ground.

c. 100

Number of impacts per match (including tackles, rucks, etc.) for a back

10.4 s

Bryan Habana's time
for 100 m

148

Average number of
tackles per game in Super
Rugby League in 2011

159

Average number of
tackles per game by
Australian test team
in 2010

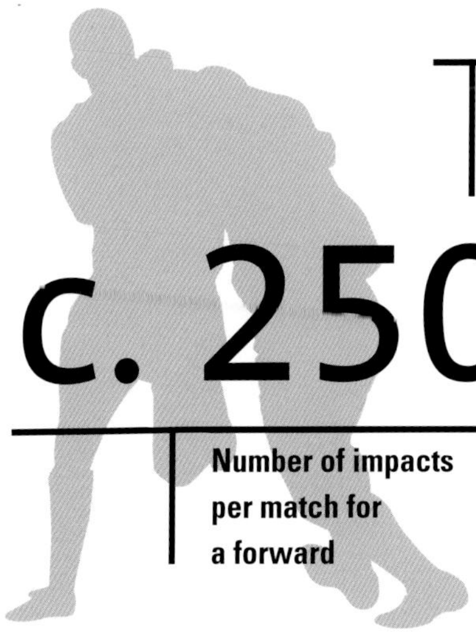

c. 250

Number of impacts
per match for
a forward

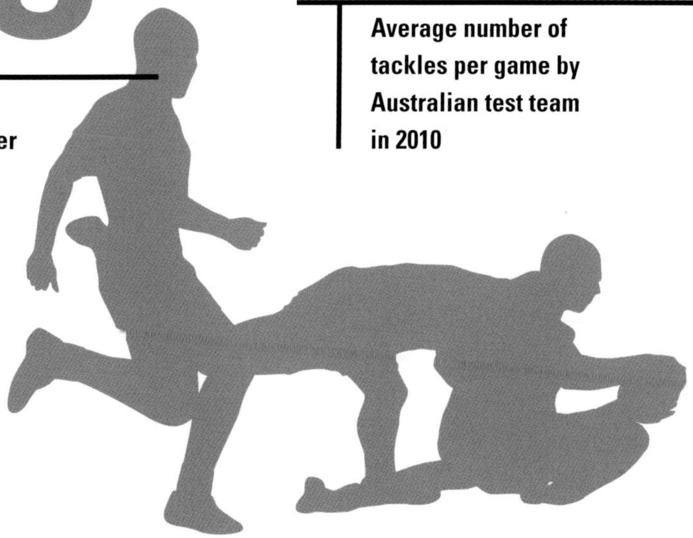

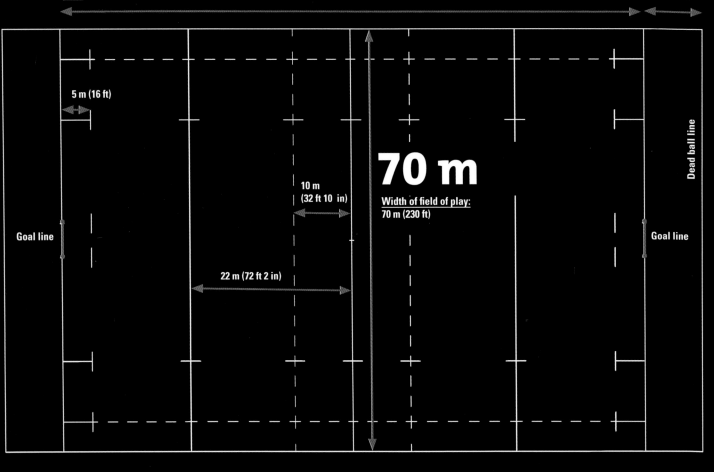

100 m

Length of field of play: 100 m (328 ft)

In-goal: not exceeding 22 m

5 m (16 ft)

70 m

Width of field of play: 70 m (230 ft)

10 m (32 ft 10 in)

22 m (72 ft 2 in)

Goal line

Goal line

Dead ball line

Game time: Two 40-min halfs with a five-min halftime and no time-outs. Any time lost due to tending injured players on the pitch is added to the end of each half

Players: Fifteen players on each team with 7 substitutions, 8 forwards and 7 backs

Points: Try: 5
Conversion: 2
Drop goal: 3
Penalty kick: 3

139

Caps won by George Gregan of Australia, the most capped rugby international of all time

1,385

Points scored by Dan Carter of New Zealand, current highest points scorer in international rugby, as of 2012

69

Most international tries, by Daisuke Ohata of Japan, who won 58 caps

1:19

Ohata's ratio of tries to games, by a colossal margin the best ratio of any international who has scored more than 30 tries

Spin the Wheel:
Cycling and Gyroscopes

19,200 rpm

Rotation speed of Hubble gyroscope flywheels

6

Number of gyroscopes in Hubble space telescope

12

Number of gyroscopes in a typical airplane

5

Number of gyroscopes in a Segway

A gyroscope is a device that maintains its orientation because it is spinning. The spinning mass has angular momentum, and because of the law of conservation of angular momentum, it resists a force applied to it to make it tip over. This can be demonstrated with a spinning bicycle wheel. If it is held upright by its axle and spun fast, it will be very hard to turn the wheel. If the axle is suspended from a string on only one end, the spinning wheel will not topple over under the influence of gravity, as might be expected, but will turn around the axis represented by the string, while still remaining vertical. This is called precession.

These counterintuitive gyroscopic effects have traditionally been invoked to explain how a bicycle stays upright when you are riding it, but in fact gyroscopic effects have little to do with bicycle stability. This is because bicycle wheels are too light and usually spin too slowly to produce much precession force. In fact you can balance on your bicycle because you quickly learn to turn the handlebars in the direction of topple, and this tends to bring the bicycle wheel back in line with the direction of movement. The faster you are going the less you need to turn the wheel to correct your balance, but when you slow down you need bigger movements, which is why it is much harder to balance a near-stationary bike.

Gyroscopic motion can keep a riderless bike upright if it is traveling fast enough and is launched in a straight line to begin with, and it may also help those who ride without using their hands.

MATHEMATICS OF A TYPICAL BIKE WHEEL

26 in

Diameter of a typical bike wheel

2.2 lb

Peripheral mass of wheel

220 lb

Combined mass of bike and typical rider

6 ft 3 in

Circumference of a typical bike wheel

0.8 in

Maximum wobble allowed if bike is to be balanced by gyroscopic effect alone

12 mph

Speed of a fast rider

0.1 kg/m^2

Moment of inertia of wheel

3 rps

Spin rate of wheel

20^c

Radians per second

The Sweet Spot: Tennis, Baseball and Elastic Collisions

Baseball bats and tennis rackets both have sweet spots—points that give a particularly "clean" contact with the ball and impart maximum power and speed to the shot. In fact, there are three points on a bat or racket: the center of percussion, the node and the power spot.

The center of percussion is the center of gravity or mass: this is the spot where none of the force from a ball hitting the bat/racket creates torque. Imagine a ball hitting a bat suspended from string. If the ball strikes the bat away from the center of mass, a torque or force of rotation will be created, and the bat will twist, swing or at least wobble. If the ball hits dead on the center of mass the whole bat will move backwards, but it won't twist or swing. The exact location of the center of percussion depends on where you are holding the bat/racket, but if you hit the ball dead on at the center of percussion you won't feel it twist or wrench in your hand and you will impart momentum to the ball more efficiently, because none will be lost to torque.

When a ball strikes a bat or racket it causes it to vibrate. Vibrations are a sort of flex or wave traveling through the bat/racket, but there are two points where the amplitude of the wave is zero: these are called the nodes. One of the nodes will be on the handle, near or under your hand, so you can't hit the ball with that one, but the other one is the spot where—if you hit the ball with it—you won't feel any vibrations or jarring. You should be able to hear the difference too.

Finally there is the power spot, which produces the highest coefficient of restitution (COR) for the ball. The COR is a measure of the elasticity of the collision between the ball and a stationary bat/racket. A COR of 1 shows a perfectly elastic collision in which the ball bounces off without losing any energy/speed. Hit the ball with the spot on the bat that produces the highest COR and you make the maximum possible use of the incoming ball speed to generate speed on the return shot/hit.

110 mph

Speed of outgoing ball

200 mph in 1/1,000 s

Acceleration of ball

36,000 N

This force expressed in newtons

90 mph

Speed of incoming ball

1/1,000 s

Length of time bat and ball are in contact

50%

Amount by which the diameter of ball is distorted during impact

20%

Amount by which the diameter of bat is distorted during impact

8,000 lb

Force with which baseball bat strikes ball, according to Robert Adair in *The Physics of Baseball*

BASEBALL BATS

20 points

Increase in team batting averages after use of aluminium bats approved by the National Collegiate Athletic Association (NCAA) in 1974

White ash: wood used to make classic Louisville Slugger bat used by today's professional players

200%

Increase in home run production after approval of aluminium bats

50 yr

Minimum age of trees before being harvested for bat wood

6–8 months

Drying time for wood

10–12

Number of bats used by early professionals in an entire career

6–7

Number of bats used in a season by a modern MLB player

10%

Proportion of wood that makes the grade for use in professional bats

14 yr

Length of time baseball legend Joe Sewell (played 1920–1933) used a single bat

TENNIS RACKETS

13–15 oz

Typical weight of wooden tennis racket

7 oz

Weight of a light alloy/composite tennis racket

10.6 oz

Approx weight of tennis racket used by American tennis player Andy Roddick

90 hz

Natural vibration rate of a wooden racket

120–200 hz

Vibration rate of a modern racket

Club Speed: Golf, Levers and the Coriolis Effect

20%

Proportion that becomes kinetic energy of shaft

50%

Proportion of energy expended in swing that becomes kinetic energy of the club head

30%

Proportion that becomes kinetic energy of body and arms

Why do we need clubs, bats, rackets etc. in our ball games? Why don't we just hit balls with our hands? One reason is that balls tend to be harder than hands, so it would be painful, but this objection is overcome easily by wearing padded gloves. This is exactly what happens in fives, a game similar to squash in which players, wearing padded gloves, hit the ball with their hands.

The main reason most sports use a club, bat or racket, however, is that the principle of levers makes it possible to move the end of a long tool at speeds far greater than we can move our hands. Added to this, these tools generally have their weight distributed at the far end, enabling players to accelerate relatively high masses to very high speeds, thus generating considerable momentum (see next page).

In the simplest model of a golf swing, the club rotates around an axis of movement at your shoulder, describing an arc through the air as the arm holding the club rotates around the shoulder. The farther away the end of the club is from the axis of movement, the faster it has to travel to complete the arc in the same time as your arm. You can see the same principle at work if you look at a record spinning on a turntable—the outer rim moves much faster than the inner rim—a point on the outer rim has much farther to travel with each revolution, but since all parts of the record are spinning at the same rate, i.e. 33 revolutions per minute (rpm), it has the same amount of time in which to travel this distance. Greater distance/same time = higher speed.

In fact, a golf swing has two axes of rotation, because golfers cock their wrists and only straighten them at the last moment. This extra "slingshot" effect helps to boost the speed at which the club head is traveling as it hits the ball.

Hinges

164 fps
Speed of club on impact

41 fps
Approx speed of hands when club hits ball

7 oz
Approx weight of club head

x4
Amount that club is moving faster than hands

250 J
Kinetic energy of club on impact

90 fps
Approx speed of club if golfer swings without cocking wrists

25 g
Acceleration in relation to gravity

2.2
Factor by which club moves faster than hands if wrists not cocked

Hit the Ball: Golf and Conservation of Momentum

157 mph
Speed of golf ball as it leaves club

105 mph
Speed of fastest ever pitch in baseball

The distance a ball will travel when hit by a golf club or thrown by a person depends on the speed at which the ball is projected and the air/wind resistance it meets as it travels, which in turn depends on the ratio of its frontal area (the area acted on by wind resistance) to its mass. Although a golf ball is smaller than a baseball, they both have roughly equivalent frontal area to mass ratios, so a golf ball probably can't be thrown any farther than a baseball. A golf ball hit by a club, however, can travel much farther than a golf ball can be thrown, and similarly even a stationary baseball whacked by a batter can travel much farther than the same player can throw it. How come?

When a golf club hits a golf ball it does not simply transfer its speed. In fact the club head is moving at about 164 fps (50 m/s) on impact, but the golf ball takes off at 230 fps (70 m/s). It is this very high launch speed (about 50% faster than the launch speed of a ball that has been thrown) that makes the ball travel farther.

The boost in speed is a result of the principle of conservation of momentum.

The principle of conservation of momentum says that the total momentum of a system must be conserved. In our case the "system" is that of a golf ball being hit by a golf club. The total momentum of the club and ball must be the same after impact as before, according to the principle of conservation of momentum. Momentum is the product of mass x velocity. The head of the golf club is considerably heavier than the golf ball, so when it transfers most of its momentum to the ball, the ball must have the same total momentum. The golf ball cannot suddenly develop more mass, so instead it increases velocity.

0.0005 s

Duration of contact between club and ball (½ millisecond)

Force = mass x acceleration

0.046 x 140,000 = 6,440 N

0.046 kg

(1.6 oz) Mass of a golf ball

140,000 m/s²

(459,318 ft/s²) Acceleration of ball

6,440 N

Force acting on golf ball at impact => equivalent to ⅔ ton

445 ft

Record distance for a baseball throw

240 yd

Typical carry for a decent golf drive

SPORTS BY NUMBERS
Golf

THE COURSE

Diagram of the Old Course at the Royal and Ancient at St Andrews (R&A), the template for all golf courses

18

Holes at the
R&A course

22

Original number of
holes until 1764

THE HOLE

4.25 in

The four golf majors
- Masters Tournament (weekend ending 2nd Sunday in April)—hosted as an invitational by and played at Augusta National Golf Club
- U.S. Open (weekend ending with the 3rd Sunday in June)—hosted by the United States Golf Association (USGA) and played at various locations in the US
- The Open Championship (The Open; usually called the "British Open" in the US) (weekend containing the 3rd Friday in July)—hosted by The Royal and Ancient Golf Club, St Andrews (R&A) and always played on a links course at various locations in the UK
- Professional Golfers Association of America (PGA) Championship (4th weekend after The Open)—played at various locations in the US

THE BALL

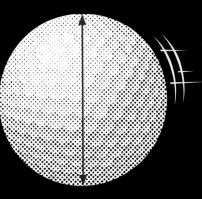

250 fps
The velocity of the ball shall not be greater than 250 fps

1.680 in
The ball must have a diameter of not less than 1.680 in

<1.62 OZ
A ball must not weigh more than 1.62 oz

THE CLUB

18–48 in
The overall length of the club must be at least 18 in and, except for putters, must not exceed 48 in

460 cc
The volume of the clubhead shall not be greater than 460 cc (28 ci)

5 in
The length of the clubhead shall not be greater than 5 in when measured from the heel to the toe

2.8 in
The height of the clubhead shall not be greater than 2.8 in when measured from the sole of the clubhead to the crown

Wins	Player	The Masters	US Open	The Open	US PGA
18	Jack Nicklaus	1963, 1965 1966, 1972	1962, 1967 1972, 1980	1966, 1970 1978	1963, 1971 1973, 1975, 1980
14	Tiger Woods	1997, 2001 2002, 2005	2000, 2002 2008	2000, 2005 2006	1999, 2000 2006, 2007
11	Walter Hagen		1914, 1919	1922, 1924 1928, 1929	1921, 1924 1925, 1926, 1927
9	Ben Hogan	1951, 1953	1948, 1950 1951, 1953	1953	1946, 1948
9	Gary Player	1961, 1974, 1978	1965	1959, 1968, 1974	1962, 1972

Left, Right, Left: The Power of the Punch

Although boxing is waning in popularity, mixed martial arts is growing at an exponential rate. All of which raises the question, who hits the hardest? The force of a punch is usually measured with some form of stress gauge, which registers the force of the impact. The "proper" scientific units for force are newtons, but most reports of punch forces give their stats in terms of pounds or kilograms. This leads to much confusion.

A blow with the force of, say, 2,200 lb, which is 1,000 kg or a metric ton, is often compared to the weight of a car, but this confuses people into thinking that being hit by this punch is like being run over by a one-ton car. In fact the car in question would be moving, and so would exert much more than a metric ton of force on impact. What the car comparison really means is that a car, or anything else that weighs a metric ton, exerts a force of 1,000 kg on the surface of the Earth under terrestrial gravity. Since a punch only lasts a fraction of a second, this "weight" only affects the punched area for an equally short time.

So a more accurate description would be saying that a 1,000 kg punch is like having a one-metric-ton car sitting on your face for a fraction of a second.

Studies show that boxing is the martial art that produces punches with the greatest force, because boxing is the one in which pugilists recruit the greatest proportion of their body mass into the punch. But even the heaviest punchers cannot achieve the force of the kicks delivered by martial artists. According to tests run by the TV show "Fight Science," the martial arts move that delivers the biggest bang of all is the Muay Thai knee kick.

1 in 5

Proportion of boxers who suffer from dementia pugilistica, the consequence of repeated blows to the skull, characterized by slurred speech and halting gait

650

Estimated number of boxers killed between 1918–1997

110 lb
Punching force of an average person with no boxing training

776 lb
Average punch of elite boxers

1,100 lb
Punching force achieved by Ricky Hatton in a test in 2007

1,000 lb
Impact force of swinging a sledgehammer into someone's face

1,420 lb
Punching force achieved by Frank Bruno in a 1985 study

53 g
Acceleration exerted on opponent's head by Bruno's punch

3.85 tons
Force claimed for punch by WBO cruiserweight champion Enzo Maccarinelli in 2007, renowned as one of hardest punchers in world boxing at the time

8,600 lb
Force of punch thrown by Ivan Drago in the film *Rocky IV*

BONES

×4

Factor by which that bone is stronger than concrete

43,000 rpm/s

Rotary acceleration that must be imparted to head for a 25% chance of knocking a person unconscious

3,300 N

Force of blow needed for 25% chance of cracking an average person's rib

9,000 N

Typical force of kicks in elite martial arts

19,000 lb

Load that one cubic inch of bone can bear—roughly the weight of five standard pickup trucks

4,000 N

Force needed to fracture the femur

ALTERNATIVE PUNCHES AND KICKS

325 lb

Average reverse punch by karate black belt

×4

Factor by which a Wushu punch is faster than a snake strike, which is between 8.2–9.8 fps

687 lb

Force needed to break a concrete slab 1 ½ in thick

1,500 lb

Force of tae kwon do spinning back kick

1,000 lb

Force of kung fu flying double kick

35 mph

Speed of car crash needed to simulate Muay Thai knee kick

Faster Than a Speeding Bullet: Pelota, Badminton and Velocity

What's the fastest ball/racket sport in the world? As we have seen, the fastest speeds are achieved by hitting quite small things with quite big things, and by using high torque and long levers to accelerate bats, rackets, clubs, etc. to very high speeds. For many years the fastest ball/racket sport in the world was confidently asserted to be the obscure Basque sport of pelota, also known as jai alai and zesta-punta.

In pelota, a small ball made of rubber and nylon and coated with goat skin (and said to be hardest ball in sports) is accelerated to a tremendous speed by slinging it out of a long wicker scoop, worn like a glove. The scoop-glove, called a cesta, allows players to generate enormous centrifugal force, hurling the ball at high velocity into a wall, whereupon a player from the other team must catch it and throw it again in a single smooth motion. The high speeds and impacts mean that the goatskin cover soon comes loose and balls must be replaced or recovered every 15–20 min.

Although the record for fastest ball in sport was long claimed by pelota player Jose Ramon Areitio with a throw made in 1979, top golf players routinely achieve comparable speeds, and in 2009 Canadian long ball champion Jason Zuback, a.k.a. "Golfzilla," was recorded hitting a ball some 16 mph faster for the TV show "Sport Science."

Yet the speed record for fastest racket sport actually belongs to the most unlikely candidate: badminton. Badminton is played with a birdie, a plug of lightweight material with a hoop of feathers. For much of play the shuttle seems to be the slowest projectile in racket sports, but with smashes some players have been timed achieving outrageous speeds.

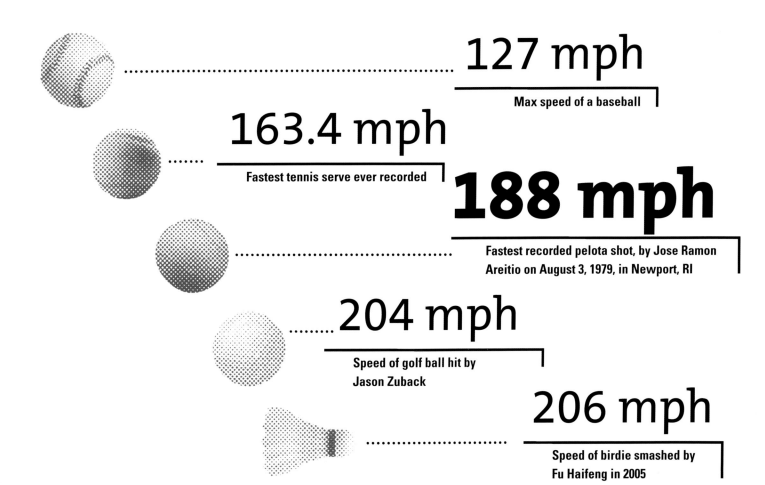

127 mph

Max speed of a baseball

163.4 mph

Fastest tennis serve ever recorded

188 mph

Fastest recorded pelota shot, by Jose Ramon Areitio on August 3, 1979, in Newport, RI

204 mph

Speed of golf ball hit by Jason Zuback

206 mph

Speed of birdie smashed by Fu Haifeng in 2005

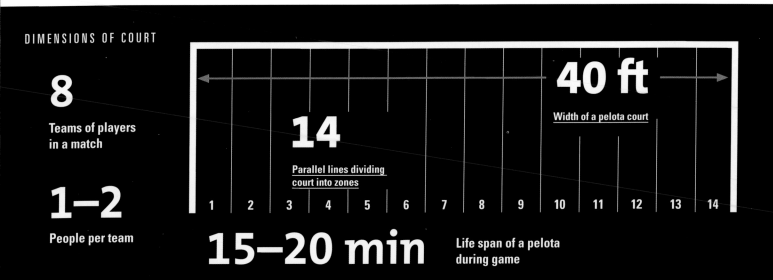

DIMENSIONS OF COURT

8

Teams of players in a match

1–2

People per team

14

Parallel lines dividing court into zones

| 1 | 2 | 3 | 4 | 5 | 6 | 7 | 8 | 9 | 10 | 11 | 12 | 13 | 14 |

40 ft

Width of a pelota court

15–20 min

Life span of a pelota during game

Gait Theory: Walking and Pendulums

A rough model for how human legs work when walking is a pendulum. A pendulum is a weight that can swing freely from a point—it converts potential energy (the energy that an object has because it has been raised in a gravitational field) into kinetic energy as it falls, and then converts kinetic energy back into potential energy on the upward swing. Pendulums are very efficient; a frictionless pendulum would swing for ever.

A pendulum-like leg increases the efficiency of walking. In fact, figures comparing actual energy consumption while walking to values predicted by a model of mammalian movement show that humans are more efficient than most mammals. Moving your leg takes energy, but using the principle of the pendulum to recapture and reuse that energy means you need to burn less to move. The precise efficiency of walking depends on a combination of speed and gait (or stride frequency and stride length). A shorter stride length is more economical at lower speeds, while at higher speeds a longer stride length is more efficient. Above a certain speed threshold—7.5 fps—running becomes a more efficient method of travel.

With human walking, the legs do not act like a simple pendulum. This would make each leg swing too slow for practical walking. By introducing extra degrees of freedom at the knee and through lifting the hips, the swing of the leg can be speeded up to achieve faster walking speeds, with the forward swing of the leg being almost entirely passive. Measurements show that there is almost no electrical activity (i.e., no muscle contraction/nerve stimulation of muscles) in the leg during the forward swing.

90 J
Energy burned if standing still

140 J
Extra energy demand from walking

200 J
Extra energy demand expected from models of general mammalian movement

4.26 fps

Walking speed for maximum efficiency

230 J

Energy burned for every meter (39 in) traveled at 1.3 m/s (4.27 fps)

300 W

Energy use of a man walking at 1.3 m/s (4.27 fps)

0.8 s

Time to swing leg forward if knee is held rigid and whole leg is swung from hip only

0.7 s

Time to swing leg forward when allowing to bend at the knee

0.6 s

Actual time of forward swing for adult walking slowly at 1.6 fps

0.35 s

Time of forward swing for adult walking fast at 6.6 fps

9.8 fps

Maximum speed at which it is possible for an adult to walk in Earth's gravity, using normal walking gait

7.5 fps

Speed at which running and walking are equally energy efficient

3.9 fps

Maximum walking speed on the Moon due to lower gravity

7.2 fps

Maximum speed at which a typical four-year-old child can walk (which is why they need to run to keep up with adults)

2.6 fps

Average walking speed of people in small villages in Greece and Crete

14.4 fps

Average walking speed for men's world record holder in 10 km walk, using rolling of hips

4.9–5.9 fps

Average walking speed of people in Brooklyn, Prague and Athens

11.8 fps

Theoretical top speed for walking with crutches

Foot Fall: Running and Load

2 million

Stress events per year on tibias of an athlete who runs 100 km/wk (62 mi/wk)

Walking may be efficient but it's murder on the feet—at least according to the measurements of the forces imposed on the human foot when walking, which are even greater while running. Standing still, you exert a downward force on the ground that acts through your feet. This is because you have mass and are in a gravity well. So your weight can be expressed in units of force (newtons) because of the force of gravity acting on your mass.

When you walk or run, your body is lifting up into the air and then landing on one foot at a time. This causes the peak load on each foot to shoot up, and it gets higher the faster you run. Sources differ on exactly how much weight/force your feet bear while running, but generally they agree that the foot of someone running slowly can experience forces of up to three times their body weight with each stride, while someone running fast may experience up to seven times their body weight.

The real problem with these high loads is not their one-time intensity, but the cumulative effect of the hundreds and thousands of steps we take each day.

If each step you take exerts a force of, say, three times your body weight on your foot, then:

- 666 N = 68 kg (150 lb): typical body weight expressed in newtons
- 666 x 3 ~ 2,000 N (235 kg/518 lb) per step
- Assuming 1,000 steps in a day: 2,000 x 1,000 = 2 million N per day
- 2 million N ~ 235,000 kg / 518,000 lb = **259 tons**

So your feet endure a cumulative force of hundreds of tons a day, bearing in mind that 1,000 steps is a conservative estimate. This is a problem because repeated small stresses can cause minute fractures in the tibia (shin bone), which spread and join up over time until they cause what is commonly known as a stress fracture but should more properly be known as a fatigue fracture.

1,150 N

Force exerted on the ground at slow running speeds 8 fps

4,700 N

Force experienced by someone weighing 670 N (150 lb) running at high speed

2,300 N

(525 lb) Maximum force, which is equal to up to three-and-a-half times body weight

30–40 N/mm²

Tensile stress needed to break bone after 2000 cycles

130 N/mm²

Single application stress needed to break same bone

Degrees of Freedom: Racket Sports and Biomechanics

Sports involving arm movements—such as tennis and fencing, through to karate and kung fu—rely on the extraordinary range and freedom of movement that the human arm provides. This in turn is a function of the number and type of joints in the arm.

Many industrial robot arms are designed with only six degrees of freedom because six is the minimum needed to reach any position. So why do we have an extra one? The answer is that it gives the flexibility to achieve the same grasp in different ways, making it possible to circumvent obstructions and reach around things. This allows, for instance, a fencer to get around an opponent's parry and score a hit, or a judoka grappling with an opponent to apply a choke hold even when being held.

Although the seven degrees of freedom of the arm allow it a wide range of movement, the mechanics of the joints may still leave "blind spots"—areas that cannot be reached. This could have important implications in sports, for example in choosing where to hit a volley at a tennis player so that he cannot get his racket to it.

Joint	Type	Degrees of freedom
Shoulder	Ball-and-socket	3
Humerus and ulna	Hinge	1
Ulna and radius	Hinge	1
Wrist	Universal	2
		Total: 7

22

Number of distinct muscles in the human arm responsible for the degrees of freedom of the arm, ignoring shoulder blade muscles and muscles that work the hand and fingers

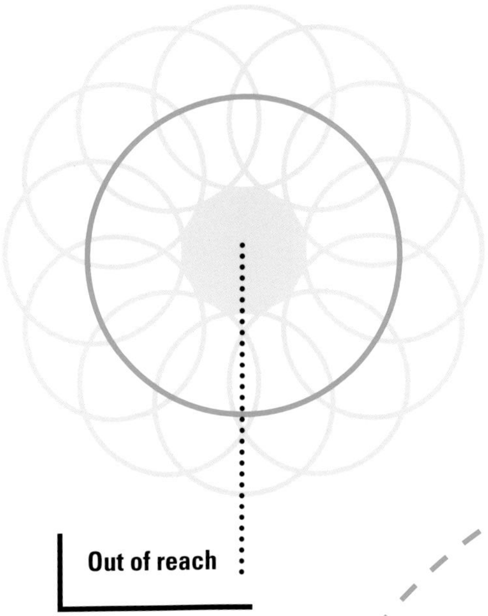

A diagram of the possible movements that an arm with two joints can make, and the resulting out-of-reach "blind spot." Although simplified, this diagram shows how human anatomy creates zones that are out of reach, which could have important implications in sports— e.g., choosing where to hit a volley at a tennis player so that he cannot get his racket to it.

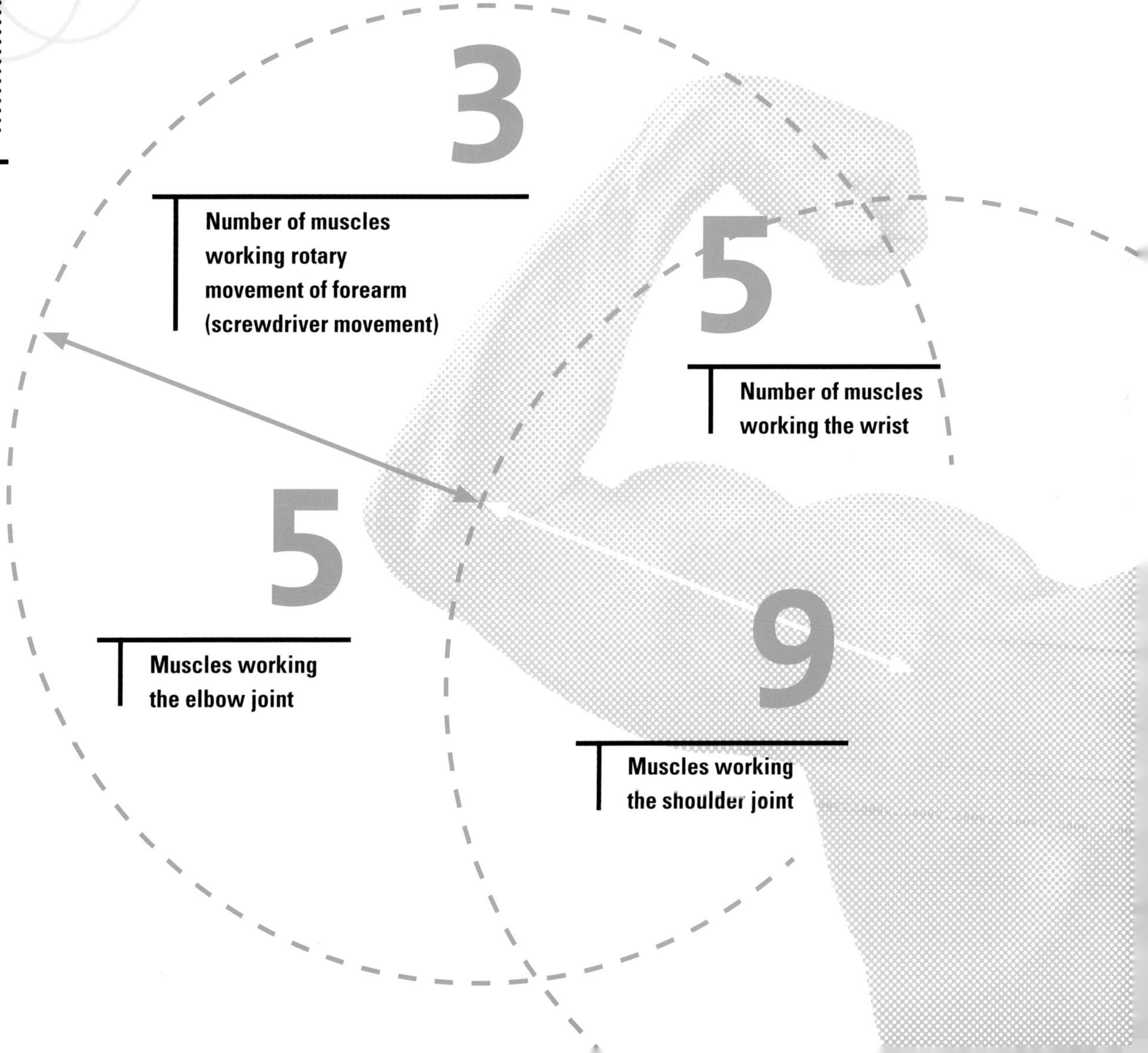

Out of reach

3
Number of muscles working rotary movement of forearm (screwdriver movement)

5
Number of muscles working the wrist

5
Muscles working the elbow joint

9
Muscles working the shoulder joint

On Board: Surfing and the Coefficient of Friction

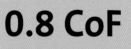

Surfers ride waves by sliding down the face on surfboards. It would be expected that the main factor governing the surfer's ability to stay on the board is his or her balance, but there is another factor to consider: the coefficient of friction (CoF). The coefficient of friction for two surfaces is a way of calculating how slippery the surfaces are when placed together, and the measurement thus governs the angle at which an object of one material will start slipping off an inclined surface of the other material: hollow wood, fiberglass or foam against human skin.

Human skin generally has quite a high coefficient of friction (it is quite sticky). Experiments in the laboratory show that the maximum tilt at which a barefoot person crouching on a wooden board can stay on before slipping is about 40°. From this, it is possible to work out that the coefficient of friction between feet and wood is about 0.8 CoF. In real conditions (i.e., with slippery seawater and boards that may be made out of other materials) the actual coefficient between a surfer and his or her board will be different. However, the lab study gives a starting point estimate, helping to work out the dividing line between staying on the board and becoming shark bait.

In actuality, the chances of making a meal for a shark are extremely slim. While surfers are indeed one of the groups most at risk from shark attack—one theory is that a surfer kicking on a board mimics the form and motions of a seal or injured sealife, which are favored shark prey—there are only between 60 and 100 shark attacks a year. The actual human death rate from shark attack is even lower than this, partly thanks to vast improvements in medical care and the speed with which victims can get medical attention. In 1900, the chances of a shark attack proving fatal were thought to be around 60%; today the chances are closer to 7%.

0.8 CoF
Between feet and wood

0.83 CoF
Between skin and a penny

0.98 CoF
Between skin and plastic

78 ft

Height of biggest wave
ever surfed, by Garret
McNamara off Nazaré,
Portugal, in 2012

8.5 stories

Equivalent height of the wave
McNamara rode

73 million

Estimated number of sharks
killed by finning each year

6 in 10

Death rate from shark
attack in 1900

1 in 4

Death rate from shark
attack today

1

Number of blue
whales stood on its
head = height of
wave surfed by
McNamara

Heavy Lifting:
Weightlifting and Gravity

Weightlifting is a sport that puts incredible stresses on the body. Ostensibly the most important factor in weightlifting is muscle strength, but the muscles do not work in isolation. They are attached to bones through tendons, and through these they exert the forces needed to overcome the weight of the loads they are attempting to lift. At the same time, stresses are transmitted through the weightlifter's body to the back and legs.

Let us return to the difference between mass and weight. The mass of an object is a measure of its inertia—how hard it is to accelerate. A large mass is just as hard to put in motion in space as it is on Earth, even though in space it is weightless. Weight is the force that gravity exerts on a mass—the larger the mass, the greater the force. To pick up a large mass in space you need to overcome its inertia, but once you've achieved this you won't need to exert any more force to keep it off the ground. On Earth you also need to overcome its inertia, but this will be comparatively insignificant in relation to

the much greater force needed to overcome the downwards force exerted by gravity.

Force = mass x acceleration. On Earth, gravity accelerates mass toward the center of the planet at 9.8 m/s^2 (32 ft/s^2). So in order to calculate in newtons (the standard unit of force) the force exerted on a mass by gravity (i.e., its weight), we need to multiply its mass in kilograms by 9.8. According to the record book, the heaviest amount lifted by a human, in the squat event in the sport of powerlifting, is 575 kg (1,268 lb). The minimum force that the powerlifter had to exert to lift this was **575 x 9.8 = 5,635 N**. This figure gives us an idea of what, in newtons, constitutes a very heavy load.

×50

Multiple of own weight ants can lift

×850

Multiple of own weight rhinoceros beetle can lift—equivalent to a human picking up an armored tank

100 N/mm²

Tensile strength of tendon

900 N

Average max amount lifted by young men bending over with their legs straight ~ 1.3 x body weight

7,200 N

(0.7 T/1,543 lb) Force compressing vertebrae of back of someone lifting 900 N

10,000 N

Average strength of lumbar vertebrae in fit young men

2,200 lb

Equivalent load a lumbar vertebra can bear without being crushed

Fast as Lightning: Sprinting and Stride Rates

6 ft 5 in
Height of Usain Bolt

5 ft 11 in
Height of Yohan Blake

6 ft 1 in
Height of Justin Gatlin

Usain Bolt is famous for his personality and records, but to sports scientists he is also a prodigy, breaking all the rules on how a sprinter should look and run. Short distance sprinting involves achieving the optimum balance between the number and length of the strides you take and the power you are able to put into each stride. A sprinter as tall as Bolt should not be able to compete, according to conventional wisdom, because with longer legs it should take him too long to get through each stride. Restricted to a lower stride rate, he should be watching shorter, more compact sprinters accelerating past him.

Even more so, extra height = extra weight, and the other tricky balance in sprinting is between having the muscle power needed to generate explosive power on the one hand, and on the other hand the problem that the bigger you are, the more power is required to move the extra mass. According to the *Journal of Sports Science & Medicine*, "The acceleration of the body is proportional to the force produced but inversely proportional to the body mass, according to Newton's second law... [which] implies an inverse relationship between height and performance in disciplines such as sprint running." A pre-Bolt era study in this journal found that the optimum height range for world champion sprinters is 5 ft 9 in—6 ft 3 in. Bolt is 6 ft 5 in tall.

But Bolt seems to have unique biomechanical characteristics that allow him to overcome the limitations of stride rate and put more power into each stride. One suggestion is that his legs are particularly good at storing and releasing energy between each stride. Having overcome the long-legs limitation, the rewards for Bolt are immense—with longer legs he can complete the course in fewer strides, allowing him to spend more time in contact with the ground with each step, which, according to the Human Performance Unit at Essex University, in turn allows him to generate force for each stride over a longer period of time.

41
Steps taken by Bolt to win Olympics 100 m final in 2012

46
Steps taken by Blake, silver medallist

0.01 s
Extra time per step that Bolt is in contact with ground

42.5
Steps taken by Gatlin, bronze medallist

4.25
Stride frequency per second of Bolt in winning 100 m

3.5
Stride frequency per second of greyhounds running at 19 m/s (62 fps)

4
Stride frequency per second of cheetah running at 29 m/s (95 fps)

Jump to It: Pole Vaulting and Conservation of Energy

31 fps

Speed of run-up of decent pole vaulters in a 1990 study

The principle of conservation of energy predicts the heights that pole vaulters can reach. Conservation of energy is similar to conservation of momentum (see page 56), in that it describes how energy cannot be destroyed or created (atomic power aside), but can only be converted from one form into another.

In the example of a pole vaulter, the vaulter builds up kinetic energy by sprinting down the track, and then plants the pole in a socket. The pole is made of very springy material, and as it bends the kinetic energy of the sprinting vaulter is converted into elastic strain energy in the bent pole. As the pole straightens back up this elastic strain energy is converted into potential energy (the energy that a mass has when elevated in a gravitational field).

We can use this sequence to predict the height of vault a pole vaulter can achieve from his speed at the end of the run-up. The vaulter has mass (M) and weight (M x G), acceleration due to gravity, runs up at speed (v) and vaults to height (H). The kinetic energy at the end of the run-up is $\frac{1}{2} Mv^2$. At the top of the vault, the vaulter is essentially moving at zero velocity, but has gained potential energy (MGH). So we can say that: $MGH \sim \frac{1}{2} Mv^2$ and $H \sim v^2/2G$. In other words, the height the vaulter will jump should be almost equal to the square of the vaulter's run-up speed over gravity x 2 (see illustration on page 77). The vaulter starts the jump standing but finishes it passing over the bar horizontally, so at the start his center of gravity is ~ 0.9 m (3 ft) off the ground but at the end his center of gravity is only about 0.1 m (4 in) over the bar.

(3 ft) Height of vaulter's center of gravity at start of jump

0.9 m

4.6 + 0.9 – 0.1 = 5.4 m

(17 ft 8.6 in) Height of bar that vaulter should be able to clear

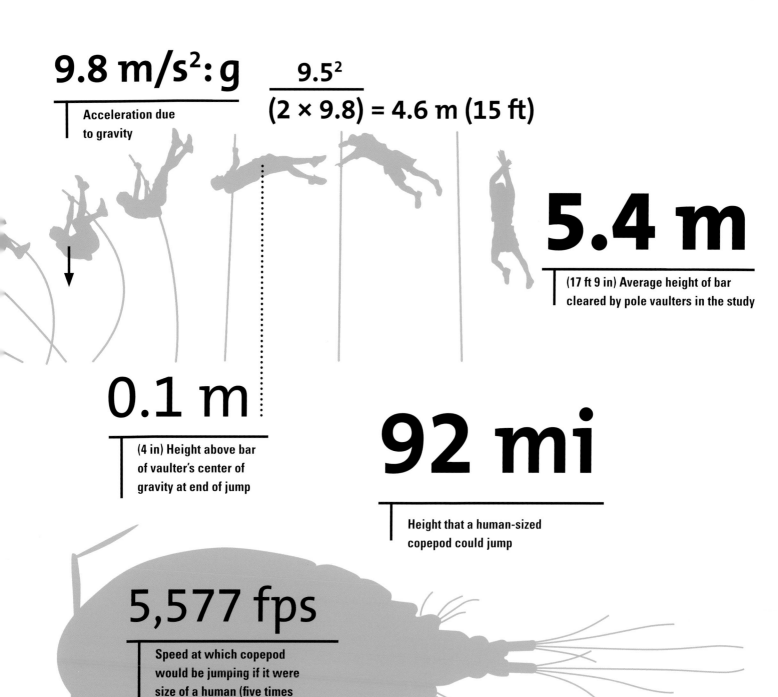

9.8 m/s²:g

Acceleration due to gravity

$$\frac{9.5^2}{(2 \times 9.8)} = 4.6 \text{ m (15 ft)}$$

5.4 m

(17 ft 9 in) Average height of bar cleared by pole vaulters in the study

0.1 m

(4 in) Height above bar of vaulter's center of gravity at end of jump

92 mi

Height that a human-sized copepod could jump

5,577 fps

Speed at which copepod would be jumping if it were size of a human (five times the speed of sound)

Hot Balls: Squash and Gas Pressure

Progression of speed and bounciness of ball

The hotter balls get, the higher they bounce. This is demonstrated in the sport of squash, where proper competitive games are impossible until players have completed a ball warm-up session. Drop a cold squash ball on the floor and it will hardly bounce at all, but smack it about for a few minutes and it becomes hot and bouncy. What is going on?

Heat makes balls bouncier by increasing the pressure inside them. Squash and tennis balls, for example, are sealed spheres filled in the middle with air. Air cannot get in or out of the sphere, so the pressure of the gas inside depends on its temperature.

The pressure of a gas is governed by the equation $p = rRT$, where "p" is pressure, "r" is density, "T" is temperature and "R" is a constant specific to the gas. Since the density of the gas in the ball cannot change, increasing the temperature must lead to increase in pressure. Conversely, lower temperatures mean lower pressure. The higher the pressure inside the ball, the bouncier it is.

In squash, the fastest ball is marked with blue dots and it is best for beginners; professionals use a ball with double yellow dots in competitions—it is the slowest and must be hit hard and continuously to stay hot. Tennis balls, on the other hand, are often stored in a fridge before tournaments to ensure that they are not too bouncy when they are brought out for play.

But why does hitting a ball make it heat up? The answer is a process called inelastic collision. So far, our discussion of the physics of bouncing balls referred to bouncing as if it were perfectly elastic—i.e., all the kinetic energy that the ball has before it collides with the floor or wall is converted back into kinetic energy, propelling the ball in the other direction. In the real world, collisions are never perfectly elastic. Some kinetic energy of the bouncing ball is dissipated as sound, and some as heat. Some of this heat goes into the ball and as the air molecules inside the ball gain energy they whiz around faster, banging harder against the inside of the ball and thus increasing pressure.

68°F

Rise in temperature
of a squash ball after
a few minutes in play

113°F

Temperature at which
squash ball reaches
heat equilibrium with
surroundings and
ceases to warm up

5 min

Length of warm-up
before match specified
by U.S. squash
governing body

2 lb

Reduction in string
tension of tennis
rackets recommended by
Tennis.com to compensate
for reduced bounce caused
by low temperatures when
playing in winter

70%

Bounce height of a frozen
golf ball compared to
ball at room temperature

134°F

Highest temperature
ever recorded on Earth's
surface, at Death Valley
in California, in 1913

Hop, Step and Spring: Hooke's Law and the Triple Jump

18.29 m

The triple jump (also known as the hop, step and jump) is a graphic illustration of a simple but overlooked principle of the mechanics of movement: that running is a form of bouncing, as are hopping and jumping. When athletes run, hop and jump, what they are really doing is bouncing along on their legs as if they were on pogo sticks. A pogo stick works because it has springs—so do our legs. The springs in your leg are tendons and ligaments.

Your tendons are elastic, which means they resist forces that are applied to them, so that when they are stretched they bounce back. As your legs move the tendons store energy by being stretched, and when they contract they release that energy to help power your movement. During running, for instance, the Achilles tendon can stretch by up to 10% of its resting length, and as it does so it stores and then releases 35% of the energy needed for each stride.

The "springiness" of a spring—i.e., how strongly it resists being stretched—is governed by Hooke's law, a principle discovered by the great 17th-century scientist Robert Hooke. Hooke's law says that the amount a spring is stretched (x) is proportional to the force applied (F): apply twice as much force and the spring will stretch twice as far. Exactly how far is determined by the spring constant, known as k. So the formula for Hooke's law is: $F = kx$. The formula for the amount of energy stored in a spring (i.e., its potential energy) is: $PE = \frac{1}{2} kx^2$. The spring constant (k) of the Achilles tendon is around 100 N/mm, and in a young adult the tendon can extend by up to 24 mm (1 in). Using these figures we can work out how much energy the Achilles tendon stores when extended: $PE = \frac{1}{2} (100 \times 24^2) = 28.8$ kilojoules. This is roughly equivalent to the energy content of 30 AA batteries or a very small bite of a chocolate bar.

(60 ft) The world record for the triple jump, achieved by Britain's Jonathan Edwards in 1995

95%

Efficiency of tendons as energy-storing springs

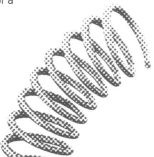

10%

Increase in tendon stiffness achieved by basketball players after an 8-week training program

2.2

The "preferred hopping frequency" per second naturally adopted by people to achieve the most efficient use of the springy tendons in the leg

10.58 m

(34 ft 8 ½ in) Distance jumped by gold-medal winner Ray Ewry of the US at the 1900 Olympic Games, when the triple jump was done from a standing start

Less is More: Physiology, Biomechanics and Training

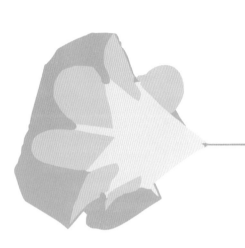

Training can be a numbers game. Like much of the rest of sports in terms of miles covered, pounds lifted, moves practised, and hours spent rehearsing, numbers rule training and they also govern the tiny percentages in performance improvement that make the difference between zero and hero. But where an appreciation of the mathematics really comes in handy is in helping to distinguish which training regimes are the most effective, and here the numbers throw up some real surprises. It turns out that sometimes less training can produce greater rewards.

For instance, an analysis in 2010 by researchers at the University of Salford in the UK showed that ice hockey players spending just 10 seconds doing resisted sprinting (where the sprinter is tethered to another player who is pulling against him) could achieve marked reduction in their sprinting speed during a game.

Another, even more striking 2009 study by researchers at the University of Copenhagen showed that by training 25% less, while including specific exercises, runners could smash their personal bests after just six to nine weeks. A study by the same researchers in 2012 showed even more dramatic results using the 10–20–30 regime, where participants spent five consecutive minutes running at low, moderate and maximal intensity for 30, 20 and 10 seconds respectively, repeating this three to four times after an initial low intensity 1 km (0.62 mi) warm-up run. Despite a 50% reduction in total training volume, the athletes in the study shaved almost a minute off their personal best times over 5 km (3.1 mi), after just seven weeks of training! Paying close attention to the numbers can help athletes to make greater performance gains while doing less training.

25%

Reduction in training volume of runners in 2009 Copenhagen study, who adopted speed endurance training

30 s

Duration of sprints in speed endurance training

6–12

Number of sprints in speed endurance training

20 min

Total length of daily training session

50%

Reduction in total amount of training by participants in the 10–20–30 Copenhagen study

0.09 s

Improvement in average 25 m sprint times thanks to resistance training, from 3.950 s to 3.859 s

2.6%

Improvement in 25 m sprint time

1 min

Improvement in average 10 km running times thanks to new training regime, from 37.3 min to 36.3 min 6/12: participating runners who obtained a new personal record

1 min

Average performance improvement of participants over 5 km

Two Wheels Good:
Cycling and Energy Efficiency

1 billion

Bicycles in the world

Since cycling improves physical fitness, cutting down on medical bills and sick days; has zero emissions and causes no pollution; and is powered by food (a renewable resource), there can be little doubt that cycling is the greenest and most desirable mode of transport available.

According to the environmental think tank the Sightline Institute, cycling is "The most energy-efficient form of travel ever devised." In terms of energy expended per unit of distance covered, it is superior to everything from walking and running to driving and horse riding. Sightline proclaimed that, "Pound for pound, a person on a bicycle expends less energy than any creature or machine covering the same distance."

Why is the bicycle so efficient? Bicycles combine a number of high-efficiency mechanisms:

Firstly they have wheels, which are highly efficient at converting energy into forward motion. Compare a wheel to your legs: the wheel coasts, whereas the legs must be accelerated from rest with every step; they have relatively little friction and are aerodynamic; they don't waste energy going side to side or up and down; and they make it easy to recover energy gained going downhill for uphill stretches.

Secondly, most bicycles have chain drives, which despite being over a century old are still among the most efficient machines ever devised.

Thirdly, bicycles have gears, which make it possible to achieve the most efficient configuration for power output whatever the speed and gradient.

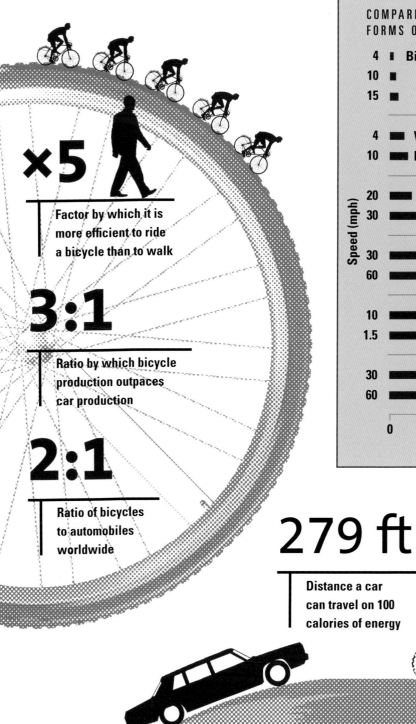

×5

Factor by which it is more efficient to ride a bicycle than to walk

3:1

Ratio by which bicycle production outpaces car production

2:1

Ratio of bicycles to automobiles worldwide

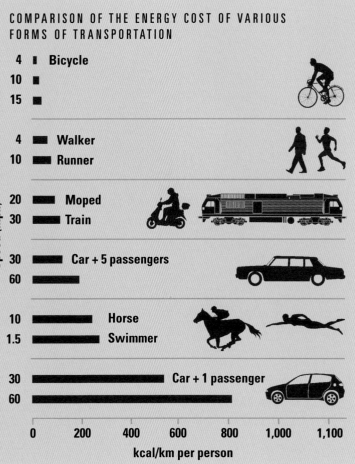

COMPARISON OF THE ENERGY COST OF VARIOUS FORMS OF TRANSPORTATION

Speed (mph)		
4	Bicycle	
10		
15		
4	Walker	
10	Runner	
20	Moped	
30	Train	
30	Car + 5 passengers	
60		
10	Horse	
1.5	Swimmer	
30	Car + 1 passenger	
60		

0   200   400   600   800   1,000   1,100

kcal/km per person

279 ft

Distance a car can travel on 100 calories of energy

3 mi

Distance a cyclist can travel on 100 calories of energy

98.6%

Efficiency achieved by chain drive in a study by engineers at Johns Hopkins University

8%

Reduction in greenhouse gas emissions if just 5% of car trips were shifted to bicycle, according to AusRoads, the association of Australian and New Zealand highway authorities

US$227 million

Annual savings to Australian public health service due to increased health of cyclists, according to a study by Melbourne University

×7

Factor by which you're more likely to be hospitalized playing Australian rules football as opposed to riding a bicycle

95%

Amount that traffic congestion is reduced when comparing a bicycle to the average car

4.3 million

Number of car trips saved per year if Sydney adopted an inner city regional bicycle network

1 lb

Amount of CO$_2$ emissions saved for every mile traveled by bike rather than by car

1 ton

CO$_2$ savings per driver if Americans substituted one-fifth of average-length car trips with bicycling

50%

Trips made that are three miles or less in the US

48 million

Equivalent number of vehicles taken off road if Americans replaced just one-fifth of average length car trips with bicycling

25%

Trips made that are less than a mile in the US

2,000%

Increase in cost of supporting a passenger kilometer (0.6 mi) of automobile traffic compared to a passenger kilometer of bicycle traffic

0.66%

Proportion of trips made in US that are by bicycle

Angles, Trajectories and Geometry

2

THE ANCIENTS CONSIDERED GEOMETRY TO BE THE SECRET STRUCTURE OF THE COSMOS, A REPOSITORY OF SACRED AND ETERNAL TRUTHS. IT CAN ALSO HELP TURN A POP FLY INTO A LINE DRIVE, OR REDIRECT A TEE SHOT FROM THE BUNKER TO THE GREEN. THIS SECTION EXPLORES ALL THINGS CURVY, SWINGING, LEANING, ANGLED AND AIMED TO SHOW THE ROLE OF GEOMETRY IN EVERYTHING FROM VOLLEYBALL TO RIDING A MOTORCYCLE UP A WALL.

The Sweetest Swing:
Mathematics of a Golf Shot

Extra distance a 7-iron shot will travel on a hot day, according to PGA Tour professional Phil Mickelson

To make a ball travel as far as possible, it should be propelled with the optimum combination of speed and height. Common sense suggests this should be at an angle of 45° from the horizontal. Any steeper and some of the impetus of the ball will be wasted in gaining altitude rather than covering horizontal distance; any shallower and the ball will hit the ground too soon, while it still had enough impetus to travel farther.

In golf, however, the attributes of the ball and club combine to make a much shallower launch trajectory more favorable and capable of achieving longer distances. Golf club heads are wedge-shaped, so that when it strikes the ball the face of the club head is angled relative to the vertical. This immediately imparts very rapid backspin or "bottom-" spin to the golf ball. The spinning ball experiences an upward Magnus force (see page 18 for more on the Magnus effect). This upward-forcing Magnus effect is enhanced by the dimples on the golf ball.

Bottom spin and dimples combine to generate such a powerful upward Magnus force that the initial flight path of the golf ball may even curve up (as opposed to the parabolic trajectory experienced by a non-spinning ball), if the ball is spinning fast enough. In other words, the lift generated by the spinning ball is enough to overcome the force of gravity acting on the mass of the golf ball. This acts to delay the ball's return to the ground under the influence of gravity, and makes it advisable to drive a ball off the tee at a shallower angle than 45°. The Magnus force on the ball will then act to give it a similar trajectory to a non-spinning ball launched at a 45° angle, although the former will not travel quite as far as the latter.

Loft is the angle of the face of the club compared to the shaft. Golf club irons vary: the higher the loft on a club, the higher the trajectory

Shaft

Loft angle

Head

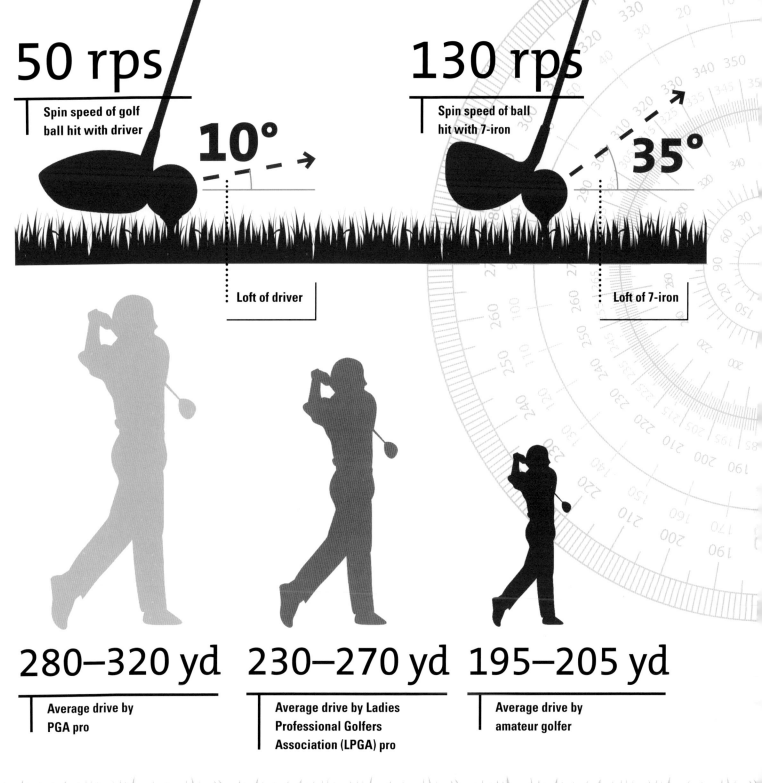

50 rps

Spin speed of golf ball hit with driver

10° →

Loft of driver

130 rps

Spin speed of ball hit with 7-iron

35°

Loft of 7-iron

280–320 yd

Average drive by PGA pro

230–270 yd

Average drive by Ladies Professional Golfers Association (LPGA) pro

195–205 yd

Average drive by amateur golfer

LOFTS AND TYPICAL DISTANCES FOR COMMON CLUBS (AMATEUR)

Club	Typical loft (in degrees)	Range for men (yd)	Range for women (yd)
Driver	8–11	200–260	150–200
3-wood	13–11	180–235	125–180
5-wood	20–22	170–210	105–170
2-iron	10	170–210	105–170
3-iron	15	160–200	100–160
4-iron	20	150–185	90–150
5-iron	25	140–170	80–140
6-iron	30	130–160	70–130
7-iron	35	120–150	65–120
8-iron	40	110–140	60–110
9-iron	45	95–130	55–95
Pitching wedge	48	80–120	50–80
Sand wedge	56	60–100	40–60
Putter	3	na	na

AMATEURS VS PROS—RANGES (IN YD) WITH COMMON IRONS

Club	Amateur	Average PGA Tour player
4-iron	170	210–220
5-iron	160	195–205
6-iron	150	180–190
7-iron	140	165–180
8-iron	130	150–170
9-iron	120	140–155
Pitching wedge	105	130–135
Sand wedge	70	115–120
Lob wedge	40	75–95

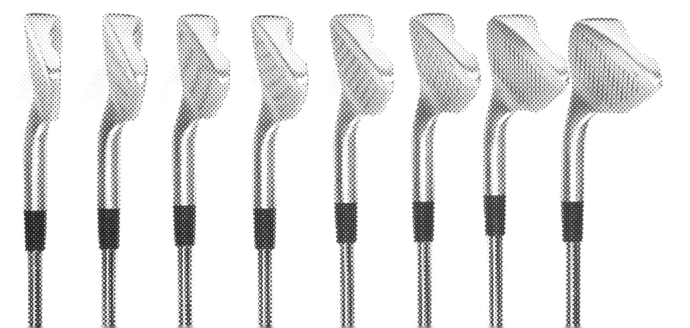

Take It to the Bank: Billiards and Geometry of Reflections

Billiards (and pool and snooker) are games of geometry at their core. Although expert players can use spin and swerve to complicate the picture, in essence billiards is about balls moving in straight lines before reflecting off planes and impacting with other points at various angles. Geometry governs the angles of the balls as they bounce off rails (cushions) and collide with one another. Even a simple working knowledge of geometry can help make you a better player.

One example is hitting a kick shot. A kick shot is where you bounce the cue ball off the rail and into a target ball. The geometry of a ball bouncing off the rail is governed by a simple law of reflection: the angle of incidence equals the angle of reflection. This means that the angle at which the ball comes off the rail is the same as the angle at which it hit the rail. Using this law you can use three simple methods to work out where to aim your kick shot.

1. HALFWAY HOUSE

If the cue ball and the target ball are the same distance away from the rail, just imagine a line between them, find the midpoint, and aim at the "image" of this midpoint on the rail. The image is the point on the rail that is exactly opposite to the object (which could be a ball, or, as in this case, a point on a line), as if the rail were a mirror and you were looking at the image reflected in the mirror.

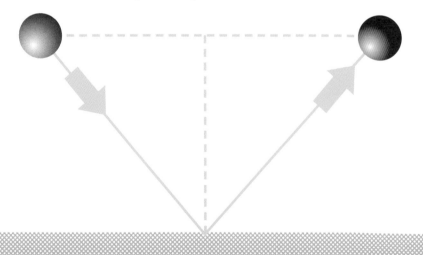

2. X METHOD

Imagine a line from the cue ball to the image of the target ball on the rail, and then a line from the target ball to the image of the cue ball. These two lines form an X—aim for the image on the rail of the intersection point of the X (i.e., where the two lines cross).

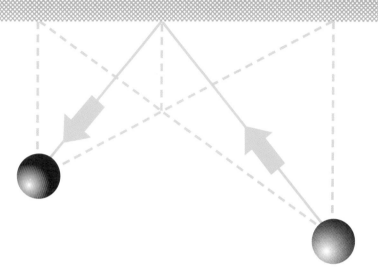

3. PARALLELS METHOD

Imagine a line between the cue ball and target ball, and find its midpoint. Now mentally draw a line from this midpoint to the image of the target ball on the rail. Now line up your shot so that it is parallel to this line.

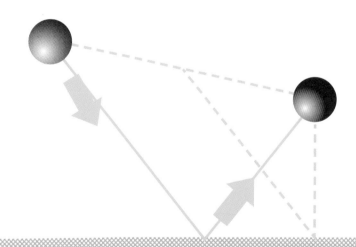

Eight Ball: Pool and Tangent Lines

Geometry also governs the outcome of a collision between two billiard balls. When a cue ball strikes another ball directly in line, almost all of the kinetic energy of the cue ball is transferred to the target ball, which continues in the same direction as the cue ball. It follows the laws of conservation of energy and momentum discussed on pages 12 and 52. So the speed of the target ball after the collision depends on the speed of the cue ball before the collision.

The two balls also follow the two laws of conservation when one strikes the other at an angle, but the outcome is more complicated because momentum is a vector quantity, which has direction as well as magnitude. Knowing the basic rule governing this collision will enable a player to work out where to aim to make the balls go in the desired directions.

After a moving ball collides with a stationary one, the second ball will proceed along a line between the centers of the two balls (the "impact line"). Meanwhile the first ball will proceed in a line perpendicular to the impact line. This new direction is along the "tangent line."

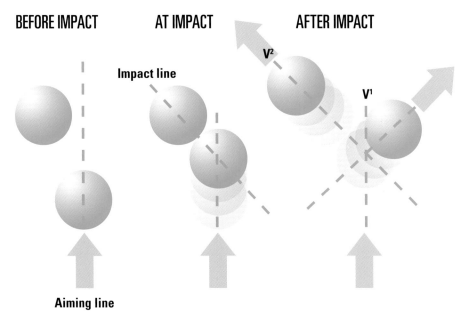

BEFORE IMPACT   **AT IMPACT**   **AFTER IMPACT**

Impact line

V^2 V^1

Aiming line

0.28 J

Kinetic energy of
cue ball in typical
billiard shot

1.8 m/s

(5.9 fps) Velocity of
cue ball in typical shot

0.17 kg

(6 oz) Mass of a
typical billiard ball

Kinetic energy (J) = $\vee$ mass (g) x velocity^2 ($\frac{m}{s}$)

$(0.5 \times 0.17) \times 1.8^2 = 0.28$ J

0.1 J

Kinetic energy of a half-
dollar coin dropped from
a height of 1 m (39.4 in)

approx 110 million trillion tons

Mass of billiard ball needed to
generate same kinetic energy as
Chicxulub impact (the crater created
65 million years ago by the asteroid
which may have helped kill off
the dinosaurs) if traveling at
normal hilliard shot speed

1.5 trillion m/s

(~ 3 trillion mph) Speed at
which normal-sized billiard
ball would need to be
propelled to generate same
same kinetic energy as
Chicxulub impact

Spiked: Mathematics of Volleyball

What does an ancient Greek philosopher have in common with a modern-day beach babe? Answer: you could use either to illustrate the mathematics of volleyball. Pythagoras is the philosopher in question, to whom is attributed the famous theorem about the sides of a right-angle triangle: the square of the hypotenuse is equal to the sum of the squares of the other two sides. In algebraic terms: $c^2 = a^2 + b^2$

What does the Pythagorean theorem have to do with volleyball? The key move in any match is the "spike." The spike is an overhead smash in which one player "sets" the ball—typically the second touch on the court playing by the standard three-touch rule—by using the tips of their fingers to push it straight up into the air near the net, and another player launches him or herself into the air to hit it hard and fast into the opposite court. The lines described by the height of the ball above the court at the "set," the path of the ball into the opposite court, and the line along the floor between them forms

a perfect right-angle triangle. This means that using the Pythagorean theorem we can work out exactly how far the ball must travel when it is spiked, and the precise angle at which the player needs to spike it.

The diagram on the next page shows us that Player A, jumping at the net, wants to plant his smash on the baseline at the back of the opposite half of the court, about 8 m (26 ft 3 in) away. Assuming he leaps 0.3 m (1 ft) above the net, he will strike the ball from a height of 2.73 m (9 ft).

Using basic trigonometry we can work out the angle ($\acute{\alpha}$) at which the volleyball player must spike the ball. The trigonometric function cosine is the ratio of adjacent/hypotenuse:

$$\text{cosine } \acute{\alpha} = \text{adj/hyp} = \frac{2.73}{8.5}$$

$$\acute{\alpha} = \text{inverse cosine } \frac{2.73}{8.5} = 71°$$

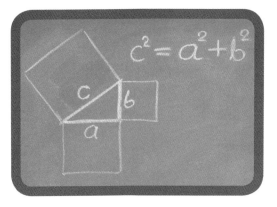

c = Hypotenuse (the square of the hypotenuse is equal to the sum of the squares of the other two sides)
b = Opposite
a = Adjacent

71°

Angle at which player must spike ball

8.5 m

(27 ft 11 in) Using the Pythagorean theorem we can see that the length of the hypotenuse, i.e., the distance the ball will have to travel, is the square root of the sum of the squares of the other two sides, i.e., $\sqrt{8^2 + 2.73^2} = 8.5$

19°

Since the angles in a triangle must add up to 180°, we can easily work out that the angle x must = $180 - (71 + 90) = 19°$

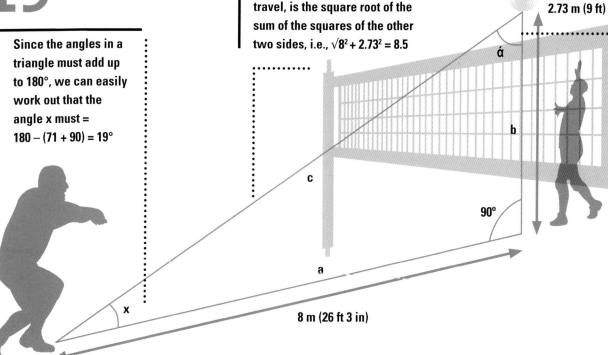

2.73 m (9 ft)

90°

8 m (26 ft 3 in)

The Racing Line:
Auto Sports and Distance

3.5 g

Sustained cornering force of a Formula 1 (F1) car

In auto racing the fastest car does not always win. The final result comes down to driving skill, and nowhere is this more apparent than in the vital art of cornering. Cornering well means maintaining maximum speed until the last possible moment, braking only to the extent absolutely necessary, maintaining relatively high speed around the corner and being able to accelerate as soon as possible coming out of the corner. Brake too late and you may hit the corner too fast, losing grip and spinning out or at least having to decelerate to save yourself. Brake too early and you lose vital momentum, giving competitors the chance to pass you. The line to follow that enables you to achieve this balancing act is the racing line, and it can be determined by geometry.

However, there is more than one racing line. Which line to take depends on what comes next and on the specifics of your car. Here we will examine two options: the geometric racing line and the late apex driving line.

The geometric line is the one that gives the widest possible arc through the corner—i.e., the line that requires you to turn the wheel the least. Where the line hits the inside of the corner is the point known as the apex. In the geometric line the apex is the geometric apex of the inside of the curve—right in the middle for a 90° bend. You can trace a geometric racing line on a diagram of the curve by using a compass and drawing a circle of constant radius that passes from the outer edge of the lead-in track, through the apex and to the outer edge of the lead-out track.

The late apex line is best when a corner is followed by a straight, since it allows you to accelerate earlier and take full advantage of the straight. You need to brake later but harder, turning in to hit the "racing apex" beyond the geometric apex—it may be out of sight, around the corner from where you start braking. This line gives a lower average speed through the corner but may be faster overall, especially for more powerful cars with better acceleration.

1.3 g

Maximum cornering force of a Ferrari Enzo

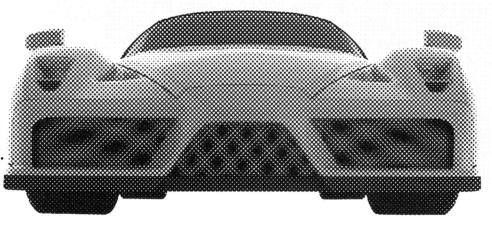

300 · 50 lb

Number of reps with a 50 lb weight that former F1 driver Juan Pablo Montoya claimed to be able to do with his neck as part of special strengthening routine for handling corners

525 lb

Load on Lewis Hamilton's body during a fast corner

193 mph

Cornering speed through turn 130R

6 g

Lateral cornering force around turn 130R at Suzuka racetrack

900 lb

Load on Lewis Hamilton's body during turn 130R

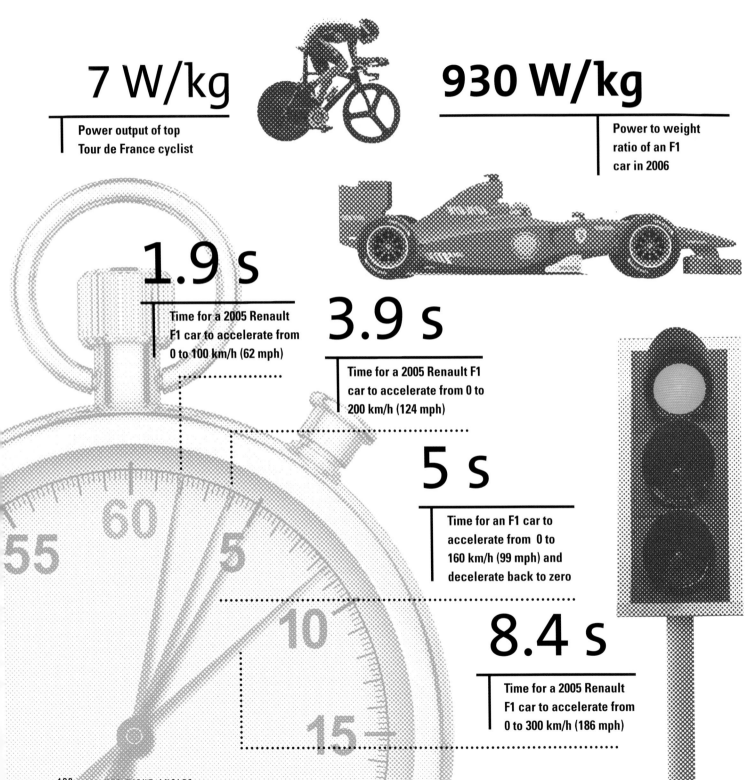

7 W/kg

Power output of top
Tour de France cyclist

930 W/kg

Power to weight
ratio of an F1
car in 2006

1.9 s

Time for a 2005 Renault
F1 car to accelerate from
0 to 100 km/h (62 mph)

3.9 s

Time for a 2005 Renault F1
car to accelerate from 0 to
200 km/h (124 mph)

5 s

Time for an F1 car to
accelerate from 0 to
160 km/h (99 mph) and
decelerate back to zero

8.4 s

Time for a 2005 Renault
F1 car to accelerate from
0 to 300 km/h (186 mph)

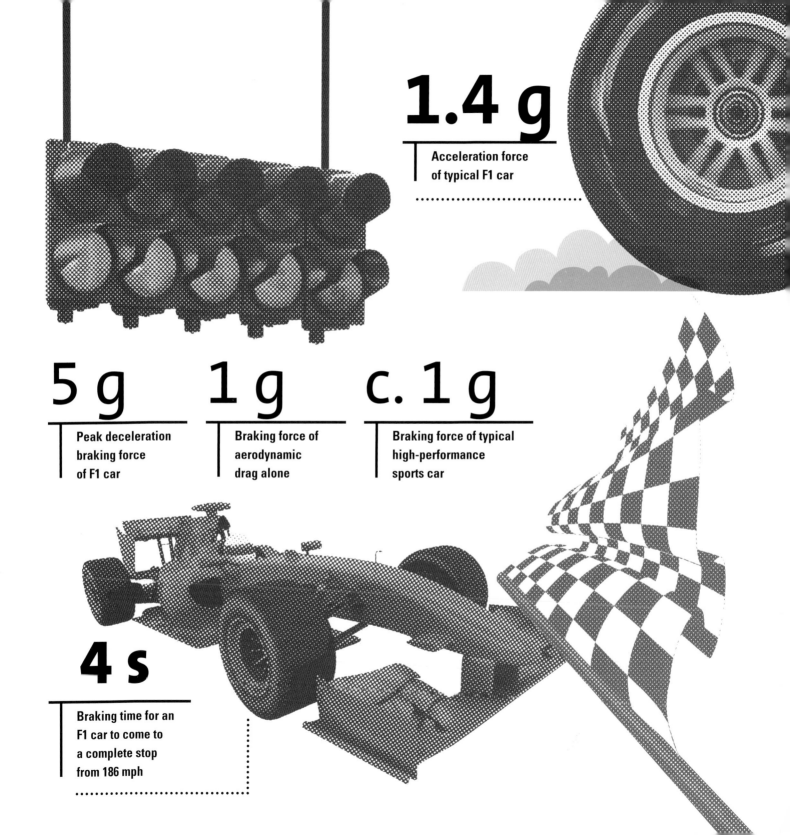

1.4 g
Acceleration force
of typical F1 car

5 g
Peak deceleration
braking force
of F1 car

1 g
Braking force of
aerodynamic
drag alone

c.1 g
Braking force of typical
high-performance
sports car

4 s
Braking time for an
F1 car to come to
a complete stop
from 186 mph

Swinging for the Fences: Baseball and Trajectories

Baseball has a dizzying array of technical terms, and that's even before slang and unofficial jargon are considered. Some of the most important terms describe types of batted balls. A batted ball is any one that the batter manages to make contact with, whether it goes fair or foul.

The three main types of batted balls are line drives, ground balls and fly balls. A fly ball is a ball that goes up in the air. There are several different types of fly balls: high fly, normal fly and pop-up or pop fly (a high fly that goes almost straight up without traveling very far horizontally). A line drive is an airborne ball with a low, flat trajectory. These tend to go very fast and are some of the hardest balls to field; they can be dangerous to pitchers, fielders and even spectators. In 2007, minor league coach Mike Coolbaugh was killed by a line drive. A ground ball is a ball that bounces or runs along the ground; these are also hard to field, and cannot be caught for an out, but, on the other hand, the chances of

a ground ball going for a home run are virtually nil at the professional level.

A line drive turns into a ground ball once it bounces. What is the dividing line between a high fly, normal fly and line drive? Ultimately the distinction is subjective, but the old formula used for working out the distance of a home run that never reached the ground—the "Tale of the Tape" (see page 106)—assigned specific values to each category, known as cotangents: line drive: 1.2; normal fly: 0.8; high fly: 0.6. The cotangent value is a ratio describing how many units of distance the ball would travel horizontally for each unit of height above the ground when the ball landed (e.g., in the uncovered stands).

1.13 runs

Average scoring rate
for a fly ball

4.3 s

Time it takes a typical fly ball
to the outfield to travel 100 yd

1.26 runs

Average scoring rate
for a line drive

4 s

Time it takes a typical line
drive to travel 100 yd

0.05 runs

Average scoring rate
for a ground ball

30%

Average number of balls in play
(fair batted balls) that go for hits,
known as batting average on
balls in play or BABIP

Tale of the Tape: Baseball, Wind Resistance and Distance

A home run is a ball that is smacked over the fence by the batter. Most Major League stadiums have tiers of seating beyond the outfield fence, so most home runs don't follow a full parabolic trajectory all the way back to the ground. Yet this does not stop TV broadcasters showing the length of a home run. How can they work out how far the home run would have traveled if it never got the chance to go the distance?

Today balls are tracked by special cameras and distance is calculated using a sophisticated 3D model of each stadium, but previously an IBM program called Tale of the Tape used spotters and subjective categorization of home run trajectories to work this out. The more sophisticated the analysis, the shorter the distance estimates have become. It used to be common to claim that epic blasts by the likes of Babe Ruth traveled 600 ft or more, but this is now deemed to be unlikely. Even the official record for longest home run ever, held by Mickey Mantle, may have been the result of an overestimate.

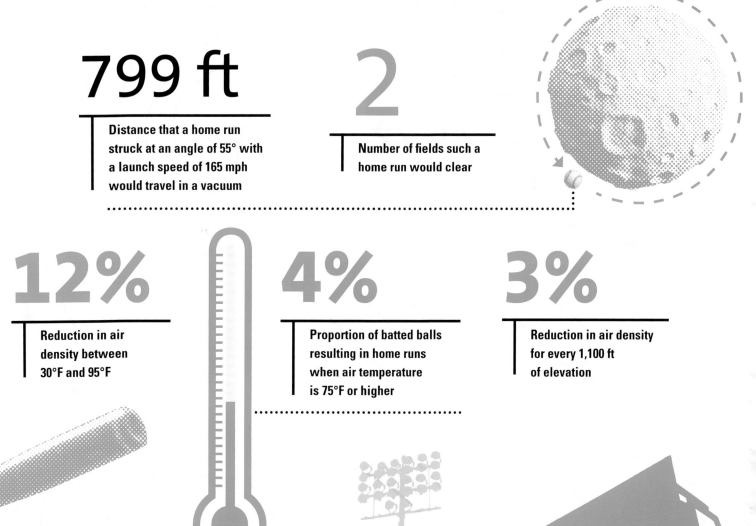

799 ft

Distance that a home run struck at an angle of 55° with a launch speed of 165 mph would travel in a vacuum

2

Number of fields such a home run would clear

12%

Reduction in air density between 30°F and 95°F

4%

Proportion of batted balls resulting in home runs when air temperature is 75°F or higher

3%

Reduction in air density for every 1,100 ft of elevation

3.2%

Proportion of batted balls resulting in home runs when air temperature is low

5,000 ft

Difference in elevation between Coors Field and Fenway Park

16%

Reduction in drag experienced by a moving baseball at Denver's Coors Field stadium vs Boston's Fenway Park

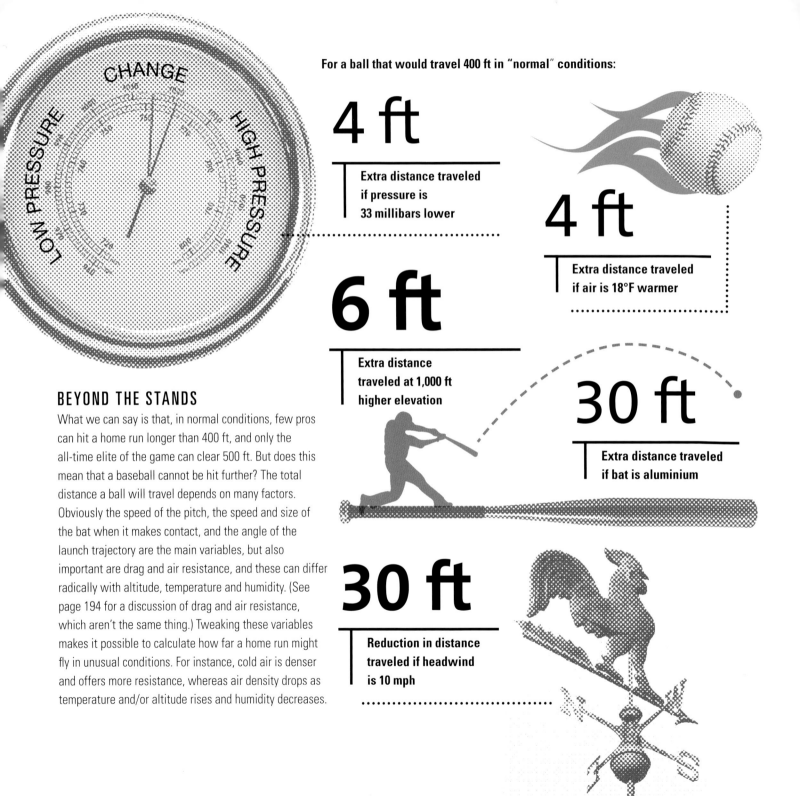

For a ball that would travel 400 ft in "normal" conditions:

4 ft
Extra distance traveled
if pressure is
33 millibars lower

4 ft
Extra distance traveled
if air is 18°F warmer

6 ft
Extra distance
traveled at 1,000 ft
higher elevation

30 ft
Extra distance traveled
if bat is aluminium

30 ft
Reduction in distance
traveled if headwind
is 10 mph

BEYOND THE STANDS

What we can say is that, in normal conditions, few pros can hit a home run longer than 400 ft, and only the all-time elite of the game can clear 500 ft. But does this mean that a baseball cannot be hit further? The total distance a ball will travel depends on many factors. Obviously the speed of the pitch, the speed and size of the bat when it makes contact, and the angle of the launch trajectory are the main variables, but also important are drag and air resistance, and these can differ radically with altitude, temperature and humidity. (See page 194 for a discussion of drag and air resistance, which aren't the same thing.) Tweaking these variables makes it possible to calculate how far a home run might fly in unusual conditions. For instance, cold air is denser and offers more resistance, whereas air density drops as temperature and/or altitude rises and humidity decreases.

Stats for the ultimate home run, assuming conditions as per a warm day at Coors Field, according to "Sport Science":

127 mph

Highest bat speed attainable by batter with perfect physique and technique

111 mph

Highest pitching speed achievable without pitcher dislocating shoulder, tearing rotator cuff or pulling a tendon off a bone

100 mph

Ball speed on impact with bat

194 mph

Speed at which ball leaves face of bat

An analysis for the show "Sport Science" has gone even further, working out the theoretical limit of distance for the ultimate home run in optimum conditions. Based on the highest possible speed of pitch that human biomechanics allow, together with the maximum bat speed achievable by the perfect human specimen timing everything to perfection and batting in the most favorable atmospheric conditions, the program came up with an estimated upper limit of 748 ft from home plate to landing point.

35°

Angle of launch trajectory

565 ft

Official record for longest ever home run, hit by Mickey Mantle at Griffith Stadium in Washington, D.C., on April 17, 1953

748 ft

Length of ultimate home run

Swing Time: Cricket and Aerodynamics

Most of the 1.5 billion people who watch cricket are aware that alongside pure speed, the most potent weapon of the bowler is swing. Swing is the ability to move the ball sideways through the air as it travels forward, creating a banana-shaped flight path when viewed from above, and deceiving the batsman into thinking he is playing in the line of the ball when he is slightly to the side or may miss it.

The key to understanding swing bowling is the concept of a boundary layer, which is when air flows past a moving ball and a thin layer clings to the surface. It can be smooth and uninterrupted (known as laminar airflow), or it can become rough and mixed up (known as turbulence). The smoother it remains the more likely it is to detach from the surface of the ball, and when this happens the air pressure increases. In swing bowling the boundary layer on one side of the ball detaches earlier than on the other side of the ball, so the air pressure is higher on one side than the other, and this causes the ball to move in the direction of the lower pressure side.

There are three kinds of swing: conventional, reverse and contrast. The first two are created by the effect that the cricket ball's seam, where the two halves are stitched together, has on the boundary layer. The seam is said to "trip" the boundary layer, which means interrupting it and setting off the mixing that turns laminar airflow turbulent. Contrast swing is caused by one side of the ball being smooth and polished, and the other being rough. (Having one side of the ball shinier than the other also helps with reverse swing.)

Which of the different types of swing a bowler can and will produce depends on factors including the age of the ball, how much work has been done polishing one side, and how fast the ball travels. Conventional swing will not happen above a certain speed threshold, while reverse swing increases with speed. Contrast swing only happens if the bowler can bowl the ball with the seam upright, as it travels, while conventional and reverse swing only happen when the ball is bowled with the seam at an angle.

70 mph
Minimum speed for contrast swing

80 mph
Speed above which there is no conventional swing

85 mph
Speed above which a new ball will start to reverse swing

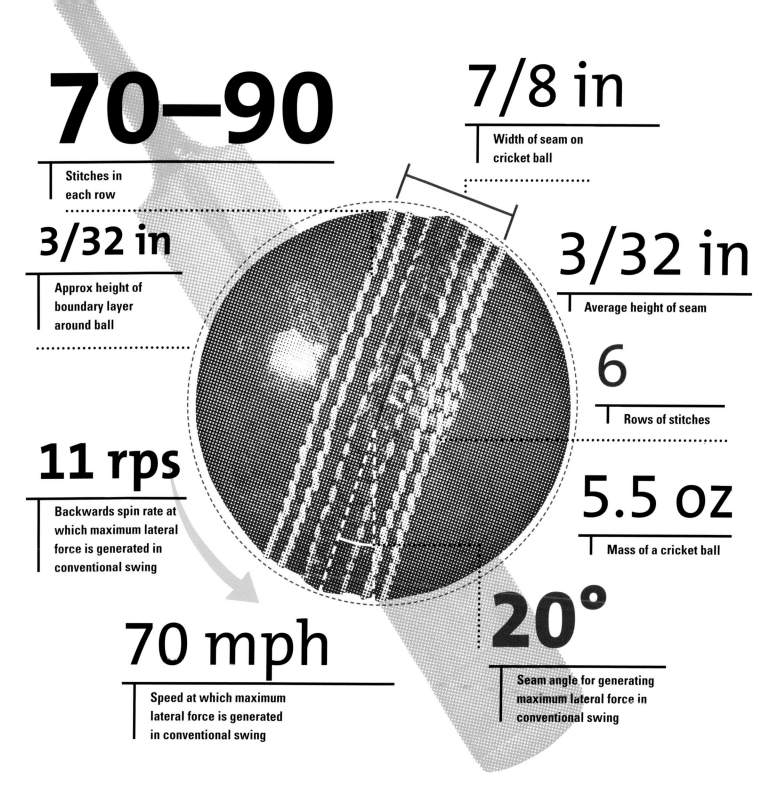

70–90
Stitches in each row

7/8 in
Width of seam on cricket ball

3/32 in
Approx height of boundary layer around ball

3/32 in
Average height of seam

6
Rows of stitches

11 rps
Backwards spin rate at which maximum lateral force is generated in conventional swing

5.5 oz
Mass of a cricket ball

70 mph
Speed at which maximum lateral force is generated in conventional swing

20°
Seam angle for generating maximum lateral force in conventional swing

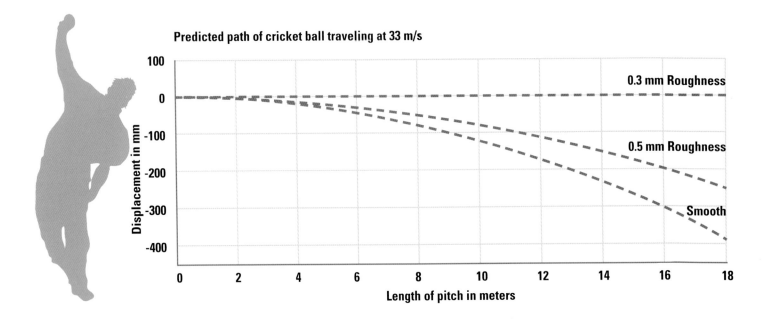

Predicted path of cricket ball traveling at 33 m/s

Displacement in mm (y-axis): 100, 0, -100, -200, -300, -400

Length of pitch in meters (x-axis): 0, 2, 4, 6, 8, 10, 12, 14, 16, 18

0.3 mm Roughness

0.5 mm Roughness

Smooth

SPIN DRIFT

Another force that can act to produce lateral movement in mid-air is the Magnus effect, produced by a spinning ball. We've already seen how this can produce curve balls in baseball and lift in golf shots (see pages 18 and 90), and the Magnus effect can also produce yet another kind of swing in cricket. Spin bowlers use special grips to impart very high revolution rates to the ball, mainly so that when it hits the ground it will bounce off at an angle, but their spinning balls also experience the Magnus effect and this makes them swing through the air before they land, although in the case of spin bowling this movement is called drift, or for a top or bottom-spinning ball, dip or loft. Similar forces are at work on a spinning soccer ball, and since soccer balls are much larger they help to illustrate the phenomenon more explicitly. One of the most famous examples in world soccer happened at a tournament in France in 1997, when the Brazilian Roberto Carlos aimed a free kick well to the right of the goal yet managed to curl it into the goal—all thanks to the Magnus effect.

80°–100°

Region on surface of seam side of ball where boundary layer separates when the ball starts to reverse swing

120°–140°

Region on surface of smooth side of ball where boundary layer separates when the ball starts to reverse swing

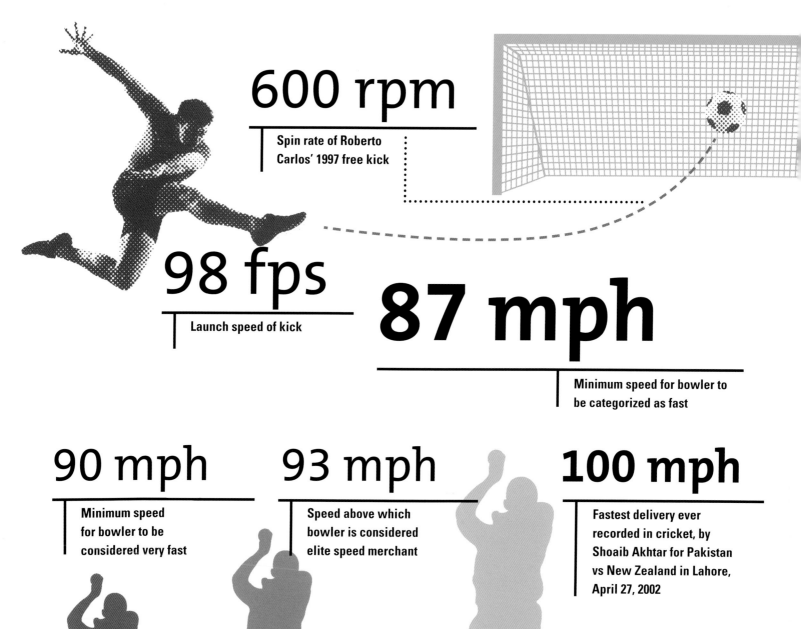

600 rpm

Spin rate of Roberto
Carlos' 1997 free kick

98 fps

Launch speed of kick

87 mph

Minimum speed for bowler to
be categorized as fast

90 mph

Minimum speed
for bowler to be
considered very fast

93 mph

Speed above which
bowler is considered
elite speed merchant

100 mph

Fastest delivery ever
recorded in cricket, by
Shoaib Akhtar for Pakistan
vs New Zealand in Lahore,
April 27, 2002

75 mph

Fastest recorded
delivery by female
cricketer, by Catherine
Fitzpatrick of Australia

Spin Class: Rotation and Bouncing in Cricket

6 ft 11 in–11 ft

Ideal range of distance
from stumps at which
spin bowler should aim

Although the spin bowler in cricket does rely to an extent on drift or dip in the flight of the ball, produced by the Magnus effect (see page 18), his main weapon is the behavior of the ball when it bounces. By imparting a high degree of spin to the ball, the bowler can make it bounce in such a way that the batsman is deceived, misses or nicks the ball, and is bowled, caught, or stumped—all ways that a batsman can be dismissed (equivalent to an "out" in baseball).

The basic science of spin bowling is straightforward. When a falling spinning object lands it will accelerate in the direction in which the top of the object is spinning (and opposite to the direction in which the bottom of the object is spinning). A simple illustration is to imagine a rotating tire dropping onto the ground. The tire will grip the surface when it lands and will gain speed in the direction in which it is spinning. A tire is effectively a two dimensional object, but a cricket ball can spin along three axes (or a combination of them): a vertical axis,

a horizontal axis pointing across the pitch, and a horizontal axis pointing down the pitch.

A ball spinning around its vertical axis may drift thanks to the Magnus effect but it won't do much on landing. A ball spinning around the crosswise horizontal axis will produce topspin or backspin, while a ball spinning around the lengthwise horizontal axis will produce lateral movement known as leg spin (spinning from right to left) or off spin (spinning from left to right). The faster the ball is spinning, the more it will be accelerated in one of these directions, but the angle of incidence (the angle at which the ball strikes the ground) is also important.

$$\frac{\text{rebound speed}}{\text{incident speed}} = \text{coefficient of restitution (COR)}$$

$$\frac{3.61}{6.26} = 0.58$$

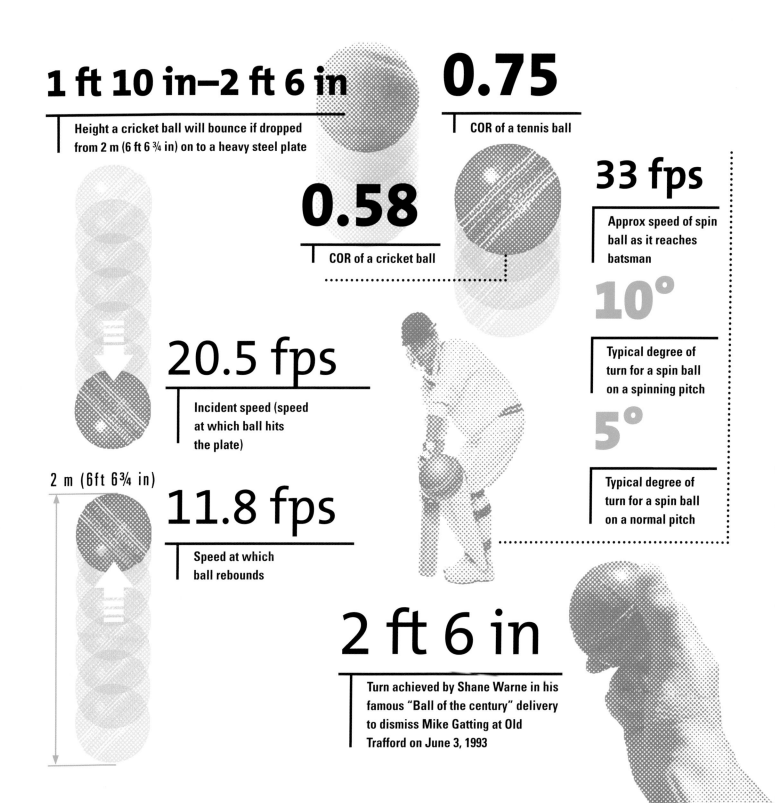

1 ft 10 in–2 ft 6 in

Height a cricket ball will bounce if dropped from 2 m (6 ft 6 ¾ in) on to a heavy steel plate

0.75

COR of a tennis ball

0.58

COR of a cricket ball

33 fps

Approx speed of spin ball as it reaches batsman

10°

Typical degree of turn for a spin ball on a spinning pitch

5°

Typical degree of turn for a spin ball on a normal pitch

20.5 fps

Incident speed (speed at which ball hits the plate)

2 m (6 ft 6¾ in)

11.8 fps

Speed at which ball rebounds

2 ft 6 in

Turn achieved by Shane Warne in his famous "Ball of the century" delivery to dismiss Mike Gatting at Old Trafford on June 3, 1993

Cricket

THE PITCH

A rectangular area 22 yd (20.12 m) in length and 10 ft (3.05 m) in width

Popping or batting crease: marked 4 ft in front of the stumps at both ends, with the stumps set along the bowling crease

Return creases: marked at right angles to the popping and bowling creases and are measured 4 ft 4 in on both sides of the middle stumps

Length of a test match:
Five days of three 2-hr sessions each, with a minimum of 90 overs to be bowled per day, with a 40-min lunch break between the first two sessions and 20 min being allowed for tea after the end of the second session

THE GROUND

A roughly elliptical field of flat grass, ranging in size from about 120–200 m (130–220 yd) across, bounded by an obvious fence, rope, or other marker

Width of field of play: 10 ft (3.05 m)

Length of pitch: 22 yd (20.12 m)

8 ft 8 in

Batting crease

Stumps

8 ft

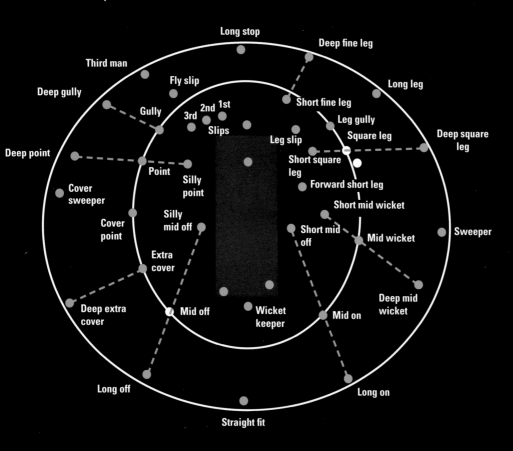

4.25 in

38 in

THE BALL

Weight: when new, not less than 5½ oz (155.9 g), nor more than 5¾ oz (163 g)
Size: not less than 8 ¹³/₁₆ in (22.4 cm), nor more than 9 in (22.9 cm) in circumference
New ball: a new ball may be requested by the fielding side after not less than 75 overs

THE BAT

Length and width:
- The overall length of the bat, when the lower portion of the handle is inserted, shall not be more than 38 in (96.5 cm)
- The width of the bat shall not exceed 4.25 in (10.8 cm) at its widest part

- Permitted coverings, repair material and toe guards, not exceeding their specified thicknesses, may be additional to the dimensions above
- Length of handle—Except for bats of size 6 and less, the handle shall not exceed 52% of the overall length of the bat

8 ¹³/₁₆ in–9 in
Circumference of ball

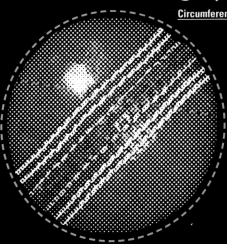

9 in
Width of all three stumps together

28 in
Height of wicket

The Wall of Death: Lean, Tilt, Gravity and Friction in Running

Force on foot

Weight

Forces acting on a sprinter accelerating

Runners on the track do not run the whole race in an upright stance. They start off out of the blocks leaning forward, and if they need to run around a bend they will lean in to it. Why do they do this, and, more importantly, does it help to answer the eternal question: could Usain Bolt run round a Wall of Death without falling off?

For a runner, forward forces must balance backwards forces, or the runner will fall over or leap into the air. Accelerating from a standing start involves exerting force almost directly backwards, so a runner must lean far forward to balance it out. A runner who tries to accelerate while standing upright will fall over.

Geometry can show how far forward the runner must lean to avoid falling over. Forces acting at an angle to each other produce a net or resultant force acting at the point where they intersect. To find the direction and strength of this resultant force, draw a parallelogram where the sides are at the angle of the forces involved

and their lengths are proportional to the magnitude of the forces. The line through the middle shows the direction and relative strength of the resultant force.

Similar considerations govern running around curves. The body must be tilted to prevent it from flying off the outside of the curve, but the top speed that can be achieved depends on the amount of friction the running surface offers. Loose surfaces such as gravel mean the runner must slow down. This problem can be overcome by banking the track. With enough speed and a large enough radius a runner could even run around a vertically tilted track, similar to a "Wall of Death"—that staple of motorcycle thrill-shows at county fairs—except for one thing: when running, both feet are off the ground at some point in a stride. A runner on the Wall of Death would simply fall straight down the first time both feet left the wall. A speed walker, however, could stick to the wall if he or she could maintain a high enough speed.

16 ft 9 in/s

Maximum speed maintained around a 12 ft radius circle by a runner who can do 33 fps on the straight

0.7

Coefficient of friction needed to prevent this runner from skidding

32 ft/s²

Approx initial acceleration of a sprinter

1.5

Coefficient of friction of rubber-soled shoes on artificial grass

12 ft 6 in/s²

Approx initial acceleration of a sports car that can do 0–60 mph in 7 seconds

0.4

Coefficient of friction of rubber soled shoes on loose surface

32 ft/s²

Acceleration of a hunting lion

16 ft/s²

Acceleration of a gazelle fleeing a hunting lion

33 ft

Minimum Wall of Death radius required for Usain Bolt to run around without falling off, if he could run with one foot on ground at all times

Launch Trajectory: Shot-Putting, Javelin-Throwing and Angles

Throwing sports go back to the ancient Olympics and even further into the past, having evolved out of the hunting techniques of the earliest humans. Before the bow and arrow were invented, spears, javelins and darts must have been the primary hunting weapons of our species. Athletes today throw everything from javelins and balls to mullets and cell phones. For them, as for their prehistoric forebears, one of the most important questions is: at what angle should the thrown object be launched to achieve the maximum distance?

The ideal launch trajectory maximizes time aloft to give the thrown object the maximum time to travel, but at the same time avoids wasting kinetic energy on gaining height that could otherwise be used to travel laterally. For a simple object that does rotate in flight, the answer is to launch it midway between the lateral and vertical components of force that must be applied—i.e., at 45°. But as soon as the thrown object starts to develop more complicated flight characteristics, this ideal launch

trajectory changes. Giving the object backspin, for instance, can result in Magnus forces that generate lift, making the ideal launch angle less than 45°.

As well as choosing the right launch angle, one way to throw farther is to increase the length of your arms. While this may not be literally possible, simple technology provides a basic but highly effective prosthetic that achieves the same result: the spear-thrower or atlatl (as it was known to the Aztecs). This is a thin wooden lath with a hollow for the butt of the javelin or dart to rest on, which dramatically increases the angular momentum imparted to the projectile so that it can be thrown much faster and farther. Similar objects are among the oldest examples of human technology ever discovered, and are believed to date back at least 70,000 years.

v^2/g

Formula for calculating how far a thrown object will travel, where v is the speed at which it is thrown and g is acceleration due to gravity

10 m

Distance an object thrown at 10 m/s (33 fps) should travel according to the formula above

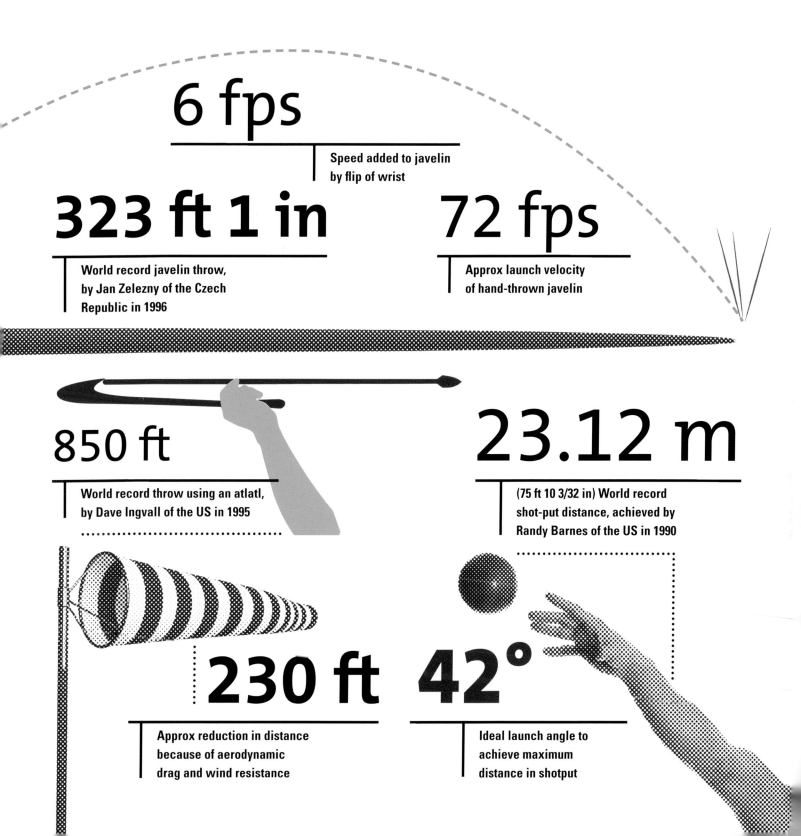

6 fps

Speed added to javelin by flip of wrist

323 ft 1 in

World record javelin throw, by Jan Zelezny of the Czech Republic in 1996

72 fps

Approx launch velocity of hand-thrown javelin

850 ft

World record throw using an atlatl, by Dave Ingvall of the US in 1995

23.12 m

(75 ft 10 3/32 in) World record shot-put distance, achieved by Randy Barnes of the US in 1990

230 ft

Approx reduction in distance because of aerodynamic drag and wind resistance

42°

Ideal launch angle to achieve maximum distance in shotput

Grade School: Cycling and Gradients

31–37 mph

The Tour de France is distinguished as a race by a number of factors: sheer length, variety of terrain and popularity. But it's the mountain stages that really impress most people. These are especially gruelling stages in which the day's racing takes the competitors up hills and even mountains, posing physical challenges quite different from racing on the flat or time-trialling.

Mountain stages consist of a series of climbs and descents. In the Tour de France and many other Grand Tours there is a different prize or jersey in contention specifically for climbing. In the Tour de France it is the Polka Dot Jersey, the winner of which is known as the King of the Mountains. This mini-competition within a competition (see page 180) is decided on points, which are awarded to cyclists for each climb in a stage based on their position (i.e., who is first up the hill) and, crucially, the category of the climb itself. In the Tour de

France the climbs are categorized from 4 (the easiest) through 1 to *hors catégorie* ("beyond classification"). The organizers could simply run a 1–5 classification system, so this last term is a conceit akin to "turning it up to 11."

Average speed of leading cyclists on a flat stage in the Tour de France

POLKA DOT JERSEY POINTS SYSTEM

25, 20, 16, 14, 12, 10, 8, 6, 4, 2

Polka Dot Jersey Points earned repectively for first 10 riders to finish *hors catégorie*

10, 8, 6, 4, 2, 1

Polka Dot Jersey Points earned repectively for first 6 riders to finish category 1

5, 3, 2, 1

Polka Dot Jersey Points earned repectively for first 4 riders to finish category 2

2, 1

Polka Dot Jersey Points earned repectively for first 2 riders to finish category 3

1

Polka Dot Jersey Points earned for first rider to finish category 4

CATEGORIES

General characteristics of the hill-climb stages in Tour de France

Category 4
Short and easy

Category 3
Approx 3.1 mi (5 km), average grade 5%, total ascent 500 ft (150 m)

Category 2
3.1 mi (5 km) or longer, 8% grade, total ascent 1,600 ft (500 m)

Category 1
12.4 mi (20 km), 6% grade, total ascent 4,921 ft (1,500 m)

Hors catégorie
>7% grade, total ascent > 3,280 ft (1,000 m)

6–12 mph

Average speed of leading cyclists on a 7.5% gradient climb

GRADIENTS

The steepness of a slope is defined by its gradient, which is the ratio of height ascended to distance traveled, often expressed as a percentage. For instance, a slope that climbs 1 m for every 100 m traveled has a 1:100 or 1% gradient. To give an idea of what counts as steep, a road with a gradient of 10% or greater will generally be signposted with "steep descent" warnings for drivers.

Some Grand Tours categorize their climbs by purely objective criteria, by multiplying the gradient of the climb in percent by its length in meters. So a climb of 5% for 5 km would score 5 x 5000 = 15,000. In the Tour de France, however, the categorization of climbs is determined by a mixture of objective and subjective considerations, including road surface, position in the stage and whether the organizers want to encourage competition (a higher category climb is worth more points so riders will fight harder to be first up it).

SIGNIFICANT CLIMBS IN THE TOUR DE FRANCE

Mont Ventoux
Average gradient: 7.1%
(Max 11%)
Length: 14.1 mi
Height gained: 5,322 ft

6,273 ft

Monte Zoncolan
From the Giro d'Italia, this climb is regarded as one of the toughest in Europe.
Average gradient: 11.9%
(Max 22%)
Length: 6.3 mi
Height gained: 3,947 ft

Alpe d'Huez
Average gradient: 8.1%
(Max 10.6%)
Length: 8.6 mi
Height gained: 3,707 ft

6,102 ft

5,676 ft

Col d'Aubisque
Average gradient: 7.0%
(Max 10.0%)
Length: 10.3 mi
Height gained: 3,809 ft

5,607 ft

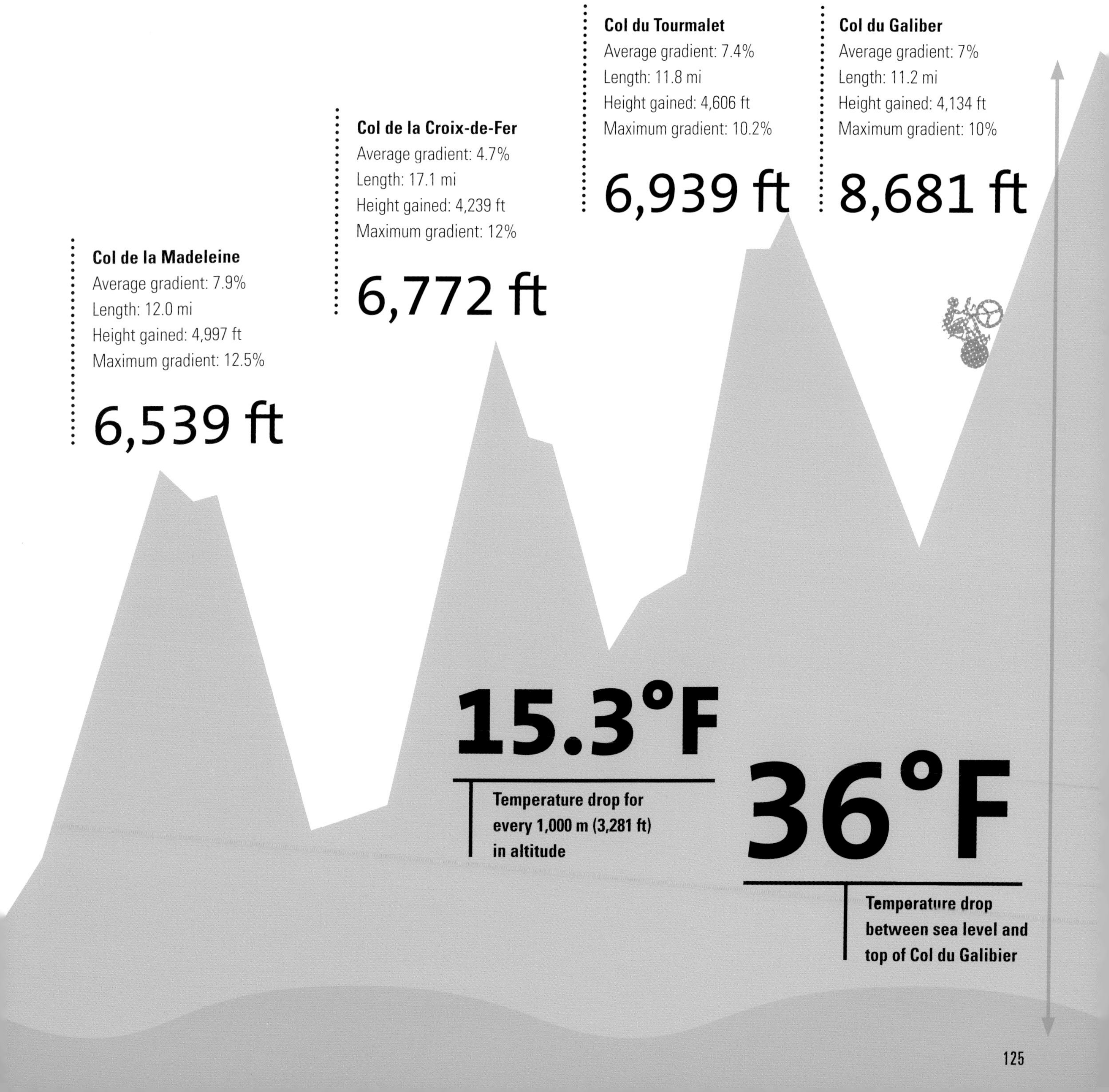

Col de la Madeleine

Average gradient: 7.9%

Length: 12.0 mi

Height gained: 4,997 ft

Maximum gradient: 12.5%

6,539 ft

Col de la Croix-de-Fer

Average gradient: 4.7%

Length: 17.1 mi

Height gained: 4,239 ft

Maximum gradient: 12%

6,772 ft

Col du Tourmalet

Average gradient: 7.4%

Length: 11.8 mi

Height gained: 4,606 ft

Maximum gradient: 10.2%

6,939 ft

Col du Galiber

Average gradient: 7%

Length: 11.2 mi

Height gained: 4,134 ft

Maximum gradient: 10%

8,681 ft

15.3°F

Temperature drop for every 1,000 m (3,281 ft) in altitude

36°F

Temperature drop between sea level and top of Col du Galibier

Number Crunching and Statistics

MOST FANS CAN NEVER COMPETE ON THE FIELD AT THE SAME LEVEL AS THEIR SPORTING IDOLS, BUT IN THE RECORD BOOKS AND STATS COMPILATIONS THEY CAN REIGN SUPREME. FOR SOMETHING AS PHYSICAL AND ACTIVE AS SPORTS, IT IS REMARKABLE HOW LARGE A NUMERICAL AND STATISTICAL FOOTPRINT IS LEFT BY ATHLETES OF ALL STRIPES. THIS SECTION EXPLORES THE STATISTICS OF SPORTS FROM POINTLESS TRIVIA TO REVEALING ANALYSIS.

Strike it Lucky: Statistics in Baseball

0.366

Highest BA of all time in MLB, among players who have played at least 1,000 games, belongs to Ty Cobb

0.96

Lowest ERA since 1900, achieved by Dutch Leonard in 1914 pitching for the Boston Red Sox

"Statistics are the lifeblood of baseball" wrote Leonard Koppett in *A Thinking Man's Guide to Baseball* (1967). No other sport is so obsessed with statistics, from IBB (Intentional Bases on Ball) to RAT (Ratio of "At-Bats" to Home Runs), and from Percentage Inherited Score to Batting Average Allowed with Runners in Scoring Position. Opinions differ as to which of these many stats are useful let alone necessary and sufficient. Yet since 1874, when the first batting average statistic appeared in print, baseball fans, players, coaches and reporters have been working them out diligently.

Two of the best known and most important stats are **BA (batting average)** and **ERA (earned run average)**. BA is the average of hits (not runs!) that a player scores divided by the number of "at-bats" it took him to score them. The picture is somewhat complicated by the fact that "at-bats" is not the same as appearances at the plate. Walks, hit-by-pitches, sacrifices and some other results do not count as "at-bats."

ERA is a bit more complicated. It is the leading measure of a pitcher's performance, and calculates the average number of Earned Runs (ER) a pitcher gives up per nine innings: (ER/innings pitched) x 9. Nine innings are equivalent to a game, so by factoring in a game's worth of innings, ERA makes it possible to compare pitchers' performances on an apples-to-apples basis. Confusion with the ERA stems from the "Earned Run" bit—only runs that are the pitcher's "fault" are counted, not those given up as the result of fielding errors, etc.

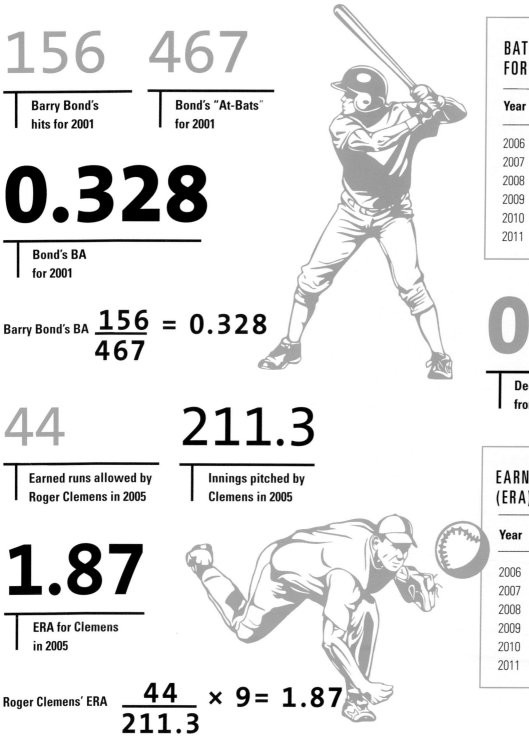

156
Barry Bond's hits for 2001

467
Bond's "At-Bats" for 2001

0.328
Bond's BA for 2001

Barry Bond's BA $\dfrac{156}{467} = 0.328$

BATTING AVERAGES (BA) FOR ALL PLAYERS IN MLB

Year	BA
2006	.269
2007	.268
2008	.264
2009	.262
2010	.257
2011	.255

0.014
Decline in BA from 2006–2011

44
Earned runs allowed by Roger Clemens in 2005

211.3
Innings pitched by Clemens in 2005

1.87
ERA for Clemens in 2005

Roger Clemens' ERA $\dfrac{44}{211.3} \times 9 = 1.87$

EARNED RUNS AVERAGES (ERA) FOR PITCHERS IN MLB

Year	ERA
2006	4.00
2007	4.91
2008	4.14
2009	4.46
2010	3.45
2011	4.52

129

Fortunate Streak: Statistics in Cricket

6/6

Number of bowlers at top of bowling average rankings who played before 1914 when conditions and rules favored bowlers more

Cricket and baseball share common ancestry as well as an obsession with statistics. In the case of cricket this is particularly understandable given that test matches can stretch over five days, during which thousands of balls will be bowled and over a thousand runs may be scored. Test matches are the long form of the game, and stats for test matches are regarded as the gold standard, although figures from both test and one-day cricket are often combined to produce simple stats like runs scored and wickets taken.

As with baseball, stats tend to break down into batting and fielding categories. For batsmen, the all-important stat is the batting average, simply known as "the average," or used as a verb (e.g., "Tendulkar averages over 50"). This is calculated by dividing the number of runs scored by the number of "outs"—not the number of innings played in, as a batsman may "carry his bat" (i.e., finish the innings undefeated). A test average of over 40 is regarded as the hallmark of a world-class batsman,

and an average of over 50 ranks a batsman among the greats of the game. It is in this context that the all-time record test average of 99.94, held by Don "The Don" Bradman, must be regarded as one of the greatest records in any sport ever (see The Bell Curve, page 136).

A major difference between stats for bowlers as opposed to batsmen is that for bowlers, lower is usually better. For bowlers the important stats include bowling average (runs conceded/wickets taken), which is equivalent to ERA in baseball, and economy rate (runs conceded per over bowled—an over consisting of six consecutive balls bowled). Both batsmen and bowlers have a statistic called strike rate, although it means different things in each case. Strike rate for batsmen shows how many runs they score per 100 balls faced, and is primarily of interest for shorter forms of the game. Strike rate for bowlers is the average number of balls bowled per wicket taken, and is potentially a better measure of the danger a bowler poses to batsmen.

113.78

One day international strike rate of Shahid "Boom-Boom" Afridi, highest for a player who has faced more than 1,000 balls

40

Test batting average regarded as the hallmark of a world-class batsman

50

Test batting average ranking batsmen as greats of the game

99.94

Test batting average of Australian cricketer Don Bradman (1908–2001)

$$\frac{\text{Economy rate}}{6} \times \text{strike rate} = \text{bowling average}$$

George Lohman's bowling average $\dfrac{1.88}{6} \times 34.1 = 10.76$

1.88

Economy rate of English cricketer George Lohmann (1886–1896)

34.1

Strike rate (number of balls bowled per wicket taken) of Lohmann

10.76

Test bowling average of Lohmann, the best for a bowler who has taken over 100 wickets and bowled more than 2,000 balls

131

Soccer Crazy: Statistics in Soccer

352

Goals scored over 518 games in Ronaldo's career

0.68

Career scoring rate of Ronaldo

14

Number of yellow cards awarded to Newcastle United player Cheick Ismael Tioté, during 2010–2011 season

Soccer is the most popular game in the world. As a lower scoring sport with less emphasis on individual performances than baseball and cricket, soccer is less obsessed with statistics. Those that matter tend to be straightforward team stats: win-draw-loss records and goal difference (goals scored minus goals conceded) being the main ones.

In the group stages of tournaments, in the event of teams being tied on points, more obscure statistics may come into play, such as corners conceded or red and yellow cards awarded. In the Euro 2012 championship, for instance, qualifiers from teams tied on points after the group stage were decided according to a mini-group of the games just between those teams, and if still tied, by criteria including total number of goals scored and position in the UEFA national team coefficient ranking system. The ranking system, which has different permutations for national teams and national leagues, is complicated enough to warrant separate discussion (see page 138).

One of the most interesting statistics for both individuals and leagues is the goals per game ratio, or scoring rate. This shows how many goals are scored, on average, per game that is played, either by an individual or in a whole league. For instance, when Brazilian striker Ronaldo Luís Nazário de Lima scored 47 goals in 49 games, as he did for Barcelona in 1996–97, his goals/game ratio was $47/49 = 0.96$. To give another example, 803 goals were scored in the 306 games played in the German Bundesliga in 2009–2010, giving a ratio of $803/306 = 2.91$.

267

Goals in 296 games
by Brian Clough

0.90

Scoring rate of Brian Clough,
highest in post-WWII
English soccer

5.27

Goals per game ratio for minor
German league Oberliga Bremen,
the league with the highest
scoring rate in world soccer

TOP CAREER SCORING RATES IN HISTORY

1.77: Fernando Peyroteo; 331 goals in 187 games
1.52: Josef Bican; 805 goals in 530 games
1.30: Imre Schlosser; 417 goals in 320 games
1.10: Franz Binder; 267 goals in 242 games
1.04: Jimmy McGrory; 410 goals in 408 games
1.04: Bernabé Ferreyra; 206 goals in 197 games
1.01: Ferenc Puskás; 754 goals in 746 games
0.93: Pelé; 767 goals in 831 games
0.93: Gerhard Müller; 735 goals in 793 games
0.90: Gunnar Nordahl; 485 goals in 537 games

GOALS/GAME RATIOS OF TOP EUROPEAN LEAGUES 2011–2012

Eredivisie (Netherlands): 3.26
Bundesliga (Germany): 2.86
Premier League (England): 2.81
La Liga (Spain): 2.76
Primeira Liga (Portugal): 2.64
Serie A (Italy): 2.56
Ligue 1 (France): 2.52

Soccer

In international matches the length of the field shall be not more than 130 yd or less than 100 yd and the width not more than 100 yd nor less than 50 yd

Lines shall not be more than 5 in in width. A flag on a pole not less than 5 ft high and having a non-pointed top, shall be placed at each corner; a similar flagpole may be placed opposite the halfway line on each side of the field of play, not less than 1 yd outside the touchline. A halfway line shall be marked across the field of play. The center of the field of play shall be indicated by a suitable mark and a circle with a 10 yd radius shall be marked round it

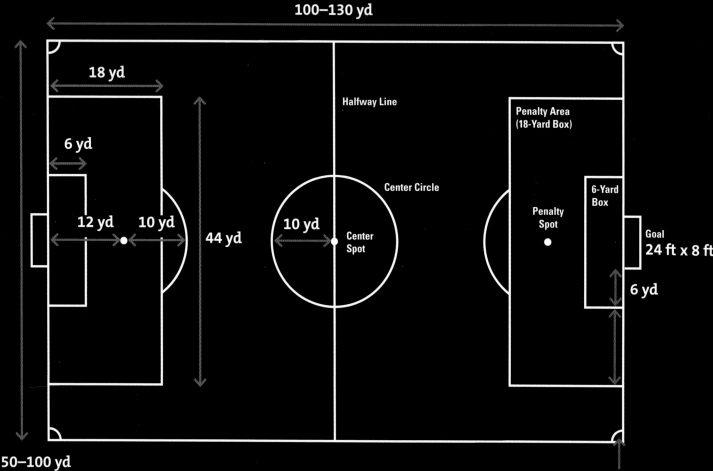

100–130 yd

18 yd

6 yd

12 yd   10 yd

44 yd

Halfway Line

Center Circle

Penalty Area
(18-Yard Box)

6-Yard
Box

Penalty
Spot

10 yd

Center
Spot

Goal
24 ft x 8 ft

6 yd

50–100 yd

1 yd radius

THE PENALTY AREA

At each end of the field of play two lines shall be drawn at right angles to the goal line, 18 yd from each goal post. These shall extend into the field of play for a distance of 18 yd and shall be joined by a line drawn parallel with the goal line. Each of the spaces enclosed by these lines and the goal line shall be called a penalty area. A suitable mark shall be made within each penalty area, 12 yd from the midpoint of the goal line, measured along an undrawn line at right angles thereto. These shall be the penalty kick marks. From each penalty kick mark an arc of a circle, having a radius of 10 yd, shall be drawn outside the penalty area

THE GOAL AREA

At each end of the field of play two lines shall be drawn at right angles to the goal line, 6 yd from each goal post. These shall extend into the field of play for a distance of 6 yd and shall be joined by a line drawn parallel with the goal line. Each of the spaces enclosed by these lines and the goal line shall be called a goal area

THE CORNER-AREA

From each corner flag post a quarter circle, having a radius of 1 yd, shall be drawn inside the field of play

THE BALL

Of a circumference of not more than 28 in (70 cm) and not less than 27 in (68 cm). Of a pressure equal to 0.6–1.1 atmosphere (600–1,100 g/cm^2) at sea level (8.5 lbs/sq in–15.6 lbs/sq in)

27–28 in

Circumference of a soccer ball

31–0

Highest scoring international soccer game ever, between Australia and American Samoa on April 11, 2001

13

Most goals scored in an international match, by Archie Thompson

16

The most goals scored in a league game, by Panagiotis Pontikos for Olympos Xylofagou vs SEK Ayios Athanasios FC in May 2007, and also by Stephan Stanis for Racing Club vs Aubry Asturies in December 1942

THE GOALS

The goals shall be placed on the center of each goal line and shall consist of two upright posts, equidistant from the corner flags and 8 yd apart (inside measurement), joined by a horizontal crossbar the lower edge of which shall be 8 ft from the ground. For safety reasons, the goals, including those which are portable, must be anchored securely to the ground. The width and depth of the crossbars shall not exceed 5 in (12 cm). The goalposts and the crossbars shall have the same width

The Bell Curve: Statistical Distribution and Performance

33

Scoring average
of Michael Jordan,
the highest in
NBA history

45

Scoring average Jordan
would need to have
achieved to be 64% higher
than next best average

If you gather data on almost any aspect of the natural world, and many aspects of the human world, including performance in sports, you will find a wide distribution of data values. For instance, if you timed 1,000 people running a kilometer, you would find some people ran it comparatively fast, some comparatively slow, and most people were somewhere in the middle. When the data is plotted on a chart or graph it will give a characteristic bell-shaped curve or "normal distribution."

Most of the data points are clustered in the middle, around what is known as the mean (average time for everyone), so the extreme ends of the curve represent times of exceptionally fast and exceptionally slow people. As a result, the bell curve is a powerful tool for analyzing performance in sports and it can be used, for instance, to predict how many people in a given population can run above a certain speed or lift a certain amount of weight.

The bell curve also provides a useful way to compare the performances of elite athletes. Distance from the middle is known as deviation, and is measured in units known as standard deviations (SDs). In a normal distribution, 68% of values are within 1 SD from the mean, 95% within 2 SDs and 99.7% within 3 SDs. So if you find a sprinter whose time for the 100 m is 4 SDs from the mean, you can tell how special his talent is. Statistically speaking, Don Bradman's test batting average (see page 137) rates as the most impressive achievement in sports, although the success of horseshoe-tossing world champion Alan Francis may be comparable.

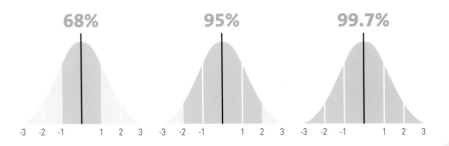

70% World class scoring rate for horseshoe tossing contestants

90% Scoring rate achieved by Alan Francis, world champion horseshoe pitcher

80% Scoring rate achievable by top two horseshoe pitchers in the world

Horseshoe tossing world championships won by Francis (out of 17)

14

99.94

Test average of Don Bradman

1

2

60.97

Test average of Graeme Pollock, the next highest after Bradman

64% Bradman's test average is 64% higher than the next highest

1 in 185,000 Expected frequency of finding a batsman who can perform at this level—in other words, one out of every 185,000 test class batsmen would be expected to have an average as high as Bradman's

Points Mean Prizes: Ranking Coefficients in Soccer

International soccer associations such as the International Federation of Association Football (FIFA) and the Union of European Football Associations (UEFA) use complex coefficient systems to rate and rank teams and leagues. These rankings form the basis of seeding for competitions, pots for qualifying group draws, number of places in major competitions such as the Champions League and even winners in the event of a tie.

Two important examples are the national team coefficient ranking system and the country coefficients. The former is a way of ranking national teams so that they can be seeded when it comes to drawing teams for the qualifying groups for major international competitions. The latter is a way of ranking national leagues to determine how many teams they can enter into the major club competitions.

In 2008, UEFA adopted a new system for determining national team coefficients. National teams earn points for each competitive match they play (10,000), with extra points for a draw (10,000) or win (30,000), more points for each goal scored (501), points lost for each goal conceded (-500) and further points for games in the later stages of tournaments (e.g., 9,000 points for a group stage game and 38,000 points for a final). Competition match points from the previous 2½ cycles are used to determine the coefficient, where a cycle consists of a qualifying round and the final tournament. The points are weighted so that the recent cycle contributes more to the score.

The country coefficient is calculated by awarding points based on participation of clubs in the Champions League and the Europa League. Clubs get points for wins and draws, and extra points for each stage of the competition they make it to. The total for each country's league are then divided by the number of teams participating, and scores for the last five seasons are combined to produce the country coefficient. These determine the ranking of each league and hence the number of clubs each gets to enter in next season's Champions and Europa Leagues.

CALCULATING COEFFICIENT FOR FRANCE IN 2006 WORLD CUP CYCLE

World Cup qualifiers	Points
France 0–0 Israel	20,000
Faroe Isl. 0–2 France	41,002
France 0–0 Ireland	20,000
Cyprus 0–2 France	41,002
France 0–0 Switzerland	20,000
Israel 1–1 France	20,001
France 3–0 Faroe Isl.	41,503
Ireland 0–1 France	40,501
Switzerland 1–1 France	20,001
France 4–0 Cyprus	42,004

Group stage	
France 0–0 Switzerland	26,000
France 1–1 Korea Rep.	26,001
Togo 0–2 France	47,002

Last 16	
Spain 1–3 France	50,003

Quarter finals	
Brazil 0–1 France	58,501

Semi finals	
Portugal 0–1 France	68,501

Final	
Italy 1–1 France	58,001

Total No. of points: 640,023
(Q. 306,014, FT- 334,009)
Total No. of matches: 17
Cycle coefficient: 37,648

UEFA 2013 COUNTRY COEFFICIENTS AS OF OCTOBER 4TH, 2012

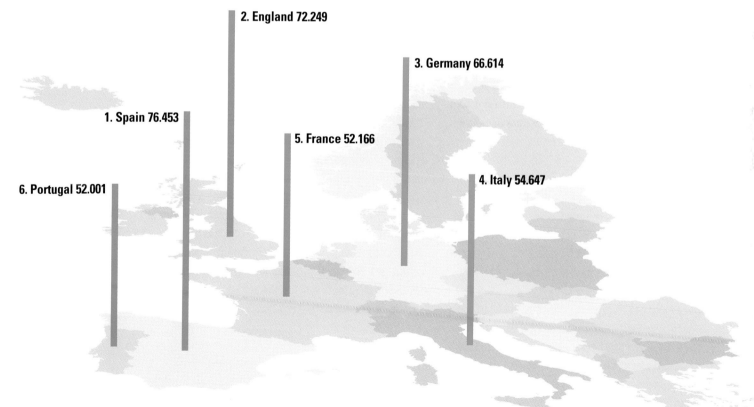

1. Spain 76.453
2. England 72.249
3. Germany 66.614
4. Italy 54.647
5. France 52.166
6. Portugal 52.001

Keeping Score:
Scoring Systems in Sport

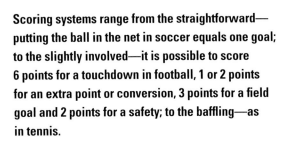

183

Number of games in the longest tennis match in history, between American John Isner and Frenchman Nicolas Mahut at Wimbledon, June 22–24, 2010

138

Number of games in the final set

Scoring systems range from the straightforward—putting the ball in the net in soccer equals one goal; to the slightly involved—it is possible to score 6 points for a touchdown in football, 1 or 2 points for an extra point or conversion, 3 points for a field goal and 2 points for a safety; to the baffling—as in tennis.

Tennis is often said to have the strangest scoring system because it includes odd terminology and seems not to be internally consistent. Scores in a tennis game proceed from "love" (zero) to 15, 30 and then 40. If the scores are tied at 40 the umpire declares "deuce." One suggestion is that scoring in increments of 15 derives from historical use of quadrants or circular scoring devices, at a time when matches consisted of six sets of four games each: (4 x 15) x 6 = 360, which in degrees makes a full circle. The third score in such a series should be 45, but it is suggested that this was contracted to 40. Love and deuce supposedly derive from the French origins of tennis: love comes from *l'oeuf* ("egg"), a descriptive

term for a zero; deuce comes from *deux*, referring to the two points needed to win from this position.

All these sports have in common, however, the objective nature of their scoring system. Scores are determined by strictly defined criteria which are either met or not met (goal-line controversy notwithstanding). Many other sports, however, have greater or lesser degrees of subjectivity in the scoring. Boxing scores, for instance, are mainly determined by successfully landing punches, but judgment about what constitutes a successful punch is subjective, and in professional boxing judges also score based on other subjective considerations, such as aggressiveness. Sports like trampolining, diving and synchronized swimming depend entirely on subjective judgments about technical and artistic merit.

11 hr, 5 min

Duration of match

8 hr, 11 min

Duration of final set

6 hr, 33 min

Previous longest match

6 hr, 31 min

Longest women's tennis match, played between Vicki Nelson and Jean Hepner in Richmond, Virginia, 1984

643

Number of shots in a single rally in this match—the longest in professional tennis

29 min

Duration of the 643 shot tennis rally

20 min

Duration of the shortest tennis match, in which Susan M. Tutt beat Marion Bandy 6–0, 6–0 at the Wimbledon tournament, 1969

HOW TO SCORE POINTS IN SYNCHRONIZED SWIMMING

- Marks are awarded by two panels of five judges each, one panel for each of two categories: technical merit and artistic impression
- Each of these categories is awarded on the basis of three considerations, such as execution, synchronization, difficulty, choreography, musical interpretation
- Marks are awarded on a scale of 10, in increments of 0.1
- For each panel, the top and bottom scores are discarded and the middle three are averaged
- The average score for technical merit is multiplied by six (since it is worth 60%)
- The artistic impression score is multiplied by four (since it is worth 40%)
- The total of the two scores constitutes the score for that routine
- Overall score for the competition is determined by scores for two routines, technical and free, which are weighted by multiplying the technical routine score by 0.35 and the free routine score by 0.65
- These scores are added to produce the final overall score for the team

Fantasy Mathematics: Mathematics and Fantasy Soccer

85%

Proportion of fantasy soccer owners who are men

Fantasy sports are games that offer participants the chance to play at being managers and/or coaches. Players assemble a fantasy team from current league rosters, working within constraints such as budget or draft order. They then manage and monitor the performance of this team through the season, with players earning points depending on their performance for each game. A fantasy soccer player would, for example, accrue points if his real-life counterpart scored a goal, kept a clean sheet or won the Most Valuable Player award.

Fantasy sports games started in the 1960s with American football aficionados in California, but soon spread to a host of other sports. Today, fantasy sports is a major entertainment industry. Fantasy football is the biggest of the fantasy sports, and its success is held responsible for driving up viewing figures and the value of broadcast rights in sports.

Fantasy sports have also acquired a second life as a teaching aid for mathematics lessons, helping to engage children and teens. This new role highlights the role of math in fantasy sports, with mathematics being involved at various levels, from simply working out your team's weekly score to complex statistical operations. In fantasy soccer, for instance, the most important part of the fantasy season is the draft, and complex models have been developed to help fantasy owners target their draft picks to produce maximum point-earning potential in the resulting team.

Mathematics have also been integrated into advanced artificial intelligence (AI) approaches to fantasy sports. An AI manager tested in the 2010/2011 English Premier League season ranked in the top 1% of fantasy soccer managers, despite not being able to deselect injured players.

41

Average age of a fantasy sports enthusiast

500

Highest ranking achieved by AI manager

26,000

Final rank of AI fantasy soccer manager at end of 2010/2011 season

98.9%

Proportion of players beaten by the AI manager

Salary of average fantasy sports enthusiast

US$50,000

38%

Proportion of students scoring A grades in mathematics tests before partipating in fantasy soccer, in a study of student athletes at Humboldt State University

83%

Proportion of the students scoring A grades in mathematics tests after participating in fantasy soccer

Faster, Higher, Stronger: Mathematics of World Records

96.5 in

World record high jump
by Javier Sotomayor

The enduring popularity of the *Guinness Book of World Records* attests to the human fascination with extremes. Sports lend themselves to extremes of human achievement, hence the motto of the modern Olympics: *Citius, Altius, Fortius*, which is Latin for "Faster, Higher, Stronger."

But how do the extremes of human achievement compare to the norms of human achievement? Comparing the record-breaking times of Usain Bolt to normal speeds is enlightening. Bolt runs so fast that he could complete the race twice and already be halfway down the third 100 m by the time an average man running at jogging pace had completed it once, and even a fast "ordinary" person would still only be at the 65 m mark as Bolt was crossing the finishing line. Bolt runs faster than a London Underground train, so he would make a good bet for commuters looking to save time.

The Olympics feature at least two disciplines in which athletes strive for height: the high jump and the pole vault. However, neither of these is very easy to compare with normal human achievement because few people have the expertise and/or guts to do a Fosbury Flop or fling themselves into the air on the end of a bendy stick. An easier metric to compare is vertical jump, and here it seems that the world record of 60 in, held by Algerian basketball player Kadour Ziani, shatters the vertical leap height you would expect from a man in the 21–30 age range, which is about 22 in (14 in for women aged 21–30). Ziani could jump over the average Filipino woman.

60 in
Record vertical leap by Kadour Ziani

48 in
Reported vertical leap of Michael Jordan

Average vertical leap for reasonably fit men aged 21–30

22 in

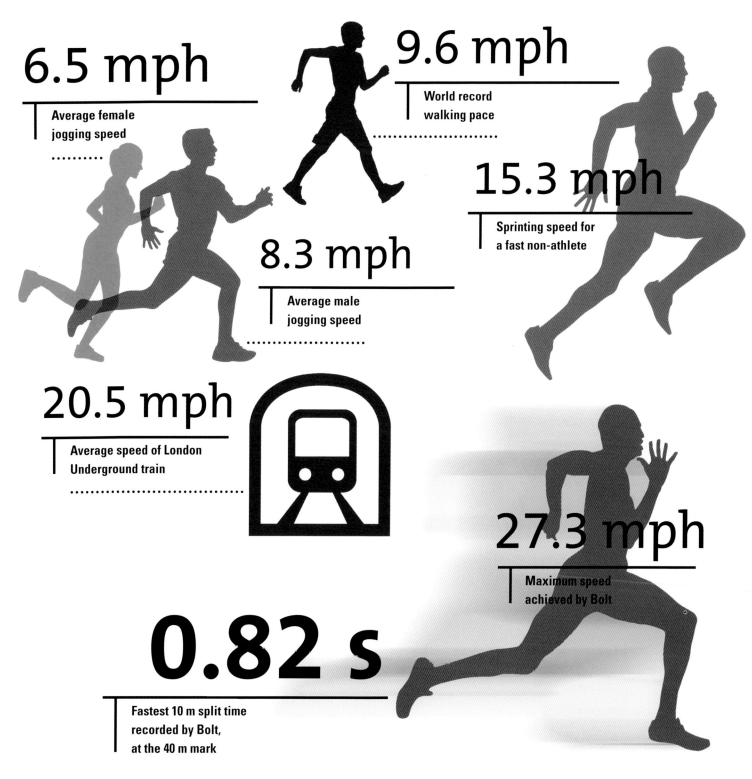

6.5 mph
Average female
jogging speed

9.6 mph
World record
walking pace

15.3 mph
Sprinting speed for
a fast non-athlete

8.3 mph
Average male
jogging speed

20.5 mph
Average speed of London
Underground train

27.3 mph
Maximum speed
achieved by Bolt

0.82 s
Fastest 10 m split time
recorded by Bolt,
at the 40 m mark

Ahead of the Curve: Geometry of World Records

2018–21

Plotting world record times against year for almost any sport and drawing a best-fit line between them will generate a downward curve. The curve starts off steep and begins to level out the nearer it gets to the present day. This is graphical evidence that humans are nearing the limits of their physical capacities in terms of athletic achievement.

For none of the sports that can be graphed, however, is the curve flat at the present. In other words the overall trend of the line is still down. Extrapolating such lines into the future tells us that records will continue to fall. With the correct application of mathematics, it could even be possible to tell when and by how much.

This is not a straightforward task, however. If you simply extend a downward curving line indefinitely it will reach zero, offering the unlikely prediction that swimmers, runners, etc., will eventually take no time at all to win their races. Complex mathematical models have to be employed using methods such as non-linear regression,

least square error logistic fit, extreme value theory and comparative modeling based on animal records.

Past attempts to predict the limits of records have not always been successful. In 1970, for instance, it was asserted that 9.60 was the ultimate threshold for human achievement in the 100 m dash but in 2009, Usain Bolt ran it in 9.58 s. Such failures have led many to take a dim view of predictions, but a recent effort before the 2012 Olympics did produce one notable success. Brian Godsey, a graduate student in mathematics at the Vienna University of Technology in Austria, used his mathematical model to predict that Kenyan David Rudisha would break his own 800 m record. He duly did so.

Years in which world records in 61 running and swimming sports will reach their limits, according to Yu Sang Chang and Seung Jin Baek of the Korea Development Institute (KDI) School of Public Policy and Management in Seoul's study in 2010

5

Events that Brian Godsey predicted might see records broken in 2012: men's 110 m hurdles and 800 m, women's 5,000 m and 3,000 m steeplechase races, women's hammer

2

Number of predictions proved right as of October 2012 (men's 800 m and 110 m hurdles)

4 min, 23 s

Maximum time that could be taken off men's marathon record, according to Mark Denny of Stanford University

68%–95%

Variability in world records over time that can be explained by population growth, according to statistician Scott Berry

2 h, 15 min, 25 s

Current world record time for women's marathon, held by Paula Radcliffe of Britain

2 h, 12 min, 41 s

Ultimate world record time for women's marathon predicted by Denny

10.19 s

Limit of record for women's 100 m dash, according to a model devised by Denny

Chance that women's 3,000 m steeplechase record will be broken in 2012, according to Godsey's model

95%

Battle of the Sexes: Gender and World Records

11 yr

Average duration of men's world records

20.43 yr

Average duration of women's world records

On average, men are bigger, taller and stronger than women. This is a biological fact, known as sexual dimorphism (body shapes that differ by sex). Human sexual dimorphism evolved at least 150,000 years ago and probably much earlier. Unfortunately the fact of sexual dimorphism has been abused for ideological and political reasons, employed to justify gender-biased stances on topics such as women in the military. In sports, however, sexual dimorphism provides a solid rationale for gender separation in many, if not most, sports.

As a result of sexual dimorphism, we would expect to find that men can run faster, jump higher and throw longer than women, and this is exactly what we find. A potential confounding factor is that although dimorphic differences are given as averages (e.g., men are 8% taller than women, on average), it is clearly the case that female athletes are way above average. Elite athletes are statistical outliers, but the same logic applies to the male athletes as well, so the gender gap can be

expected to remain. The effects of this gap can be seen in the world record (WR) times given in the table on the next page. It compares world records for men and women in 14 track-and-field events, showing the year in which the record was set and the percentage difference between the best performances in 2007–2010 and the all-time record.

The table also shows something much more controversial. World records in women's track-and-field events are almost all relatively old, and the best times/distances by modern women tend not to come close. This is almost certainly a legacy of widespread and systematic doping in the 1980s and early 1990s, which has skewed the records by pushing them beyond the physiological limits of women not on performance-enhancing drugs (PEDs). This means that in most track-and-field events today, non-doping women have no chance of breaking tainted records, which has real financial implications, since bonus prize money is awarded for records. It is likely that many of the women's records will never be broken, unless dopers can outsmart testers.

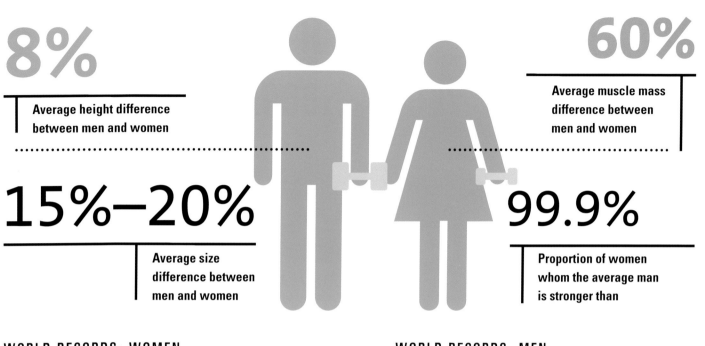

8%

Average height difference between men and women

60%

Average muscle mass difference between men and women

15%–20%

Average size difference between men and women

99.9%

Proportion of women whom the average man is stronger than

WORLD RECORDS: WOMEN

Event	WR time	Year	No. of yrs WR held	Best during 2008 –2011 vs WR
100 m	10.49	1988	24	1.43%
Short hurdles	12.21	1988	24	1.80%
200 m	21.34	1988	24	1.87%
400 m	47.6	1985	27	2.58%
400 m hurdles	52.34	2003	9	0.15%
800 m	1:53.28	1983	29	0.64%
1,500 m	3:50.46	1993	19	2.64%
5,000 m	14:11.15	2008	4	0.00%
10,000 m	29:31.78	1993	19	1.24%
Marathon	2:15:25	2003	9	3.21%
Shot put	22.63	1987	25	6.89%
Discus	76.8	1988	24	11.60%
Long jump	7.52	1988	24	5.19%
High jump	2.09	1987	25	0.48%
Average			**20.43**	**2.84%**

WORLD RECORDS: MEN

WR time	Year	No. of yrs WR held	Best during 2008 –2011 vs WR
9.58	2009	3	0%
12.80	2012	0	0%
19.19	2009	3	0%
43.18	1999	12	1.32%
46.78	1992	19	1.00%
1:40.91	2012	0	0%
03:26.0	1998	14	1.59%
12:37.35	2004	8	1.69%
26:17.53	2005	7	0.54%
2:03:38	2012	0	0%
23.12	1990	22	3.07%
74.08	1986	26	2.97%
8.95	1991	21	2.97%
2.45	1993	19	2.86%
		11	**1.24%**

The Long and the Short: Running and World Records

2:01:48

Best time possible in marathon, according to *Engineering Sport* blog

1:58:00

Best time possible in marathon, according to Dr Eddie Coyle, professor at the University of Texas

The two races that have attracted the most attention from statisticians trying to extrapolate the future limits of possible world record times are the longest and shortest ones: the 100 m dash and the marathon. In both cases efforts have been energized by landmark achievements: in the case of the 100 m, Usain Bolt's successive world records that blasted holes in previous predictions, and in the case of the marathon, the record set by Ethiopian Haile Gebrselassie in Berlin in 2008.

We will see (on page 170) how Bolt's ultimate possible time stacks up against animal Olympians. Yet this is just one of the strands used to predict how fast Bolt and other sprinters might go. His time could be boosted by perfect conditions, the perfect start, etc. Other methods include a system called extreme values theory, which uses data about less extreme events to predict the likelihood of extreme ones. The different methods give a range of predicted limits for the 100 m record. A possible confounding factor is whether or not Bolt's time reflects

the use of performance-enhancing drugs. There is no evidence for this and there are established biomechanical explanations for his amazing speed, but a look at the progression of the 100 m record over the years shows that the amounts being shaved off the top time had been progressively shrinking until Bolt's emergence. His feats are statistical outliers that have caused suspicion.

In the marathon one method used to predict the limit of possible achievement, described in the blog *Engineering Sport*, took the average difference between the average of the top 25 performances in each year and the actual top performance for the year, for every year from 1948. It found that the average difference was 2%, suggesting that the best time achievable is about 98% of the average elite performance.

2058

Year by which sub-2-hour marathon will be achieved on this trend

9.48 s

Possible time for 100 m predicted by **Mark Denny** of Stanford University

9.37 s

Possible time for 100 m predicted by **F. Péronnet** and **G. Thibault**

0.01 s

Average annual improvement on 100 m record based on average times of the top six finalists in world championship and Olympic finals since 1956, according to Professor Stephen Seiler

100 M WORLD RECORDS OVER LAST 100 YR

500 yr

Time predicted to reach 9.48 s record, according to a logistic model of times since 1912

3.08 yr

Average time between new world records

Stick or Twist?
Game Theory and Fencing

Fencing is a game of attack, defense, and counterattack. Competitors watch each other warily, waiting for their opening. Should you risk all with an extravagant lunge that could win you the match at a stroke, or should you play it safe and not offer your opponent the chance to counter? Boxing, wrestling, martial arts and many other sports involve almost identical considerations. You might think that this is the realm of psychology, but in fact there is a whole branch of mathematics that deals with this sort of dilemma. It is called game theory.

The classic illustration of game theory is the prisoner's dilemma, a mathematical demonstration showing how acting "rationally" doesn't always lead to the most rational-seeming outcome. Two criminals are arrested for a crime and interrogated in separate rooms. If both men confess to the crime, each will get 10 years in prison. If neither confess they can only be charged with a lesser offense and will each serve just one year. But if only one of them confesses, he will go free and the other man will

get 20 years. What is the most rational thing for the criminals to do? In game theory the criminals' strategies and payoffs are displayed using a "payoff table" (see box below).

		Fingers Malone	
		CONFESS	**DENY**
Mac the Knife	**CONFESS**	10 / 10	0 / 20
	DENY	20 / 0	1 / 1

If Fingers Malone thinks through his options rationally he will see that denying the crime will only pay off in one of the four possible outcomes, whereas confessing it could gain him the maximum benefit and will at least ensure he avoids getting screwed. Mac the Knife will rationalize in the same way, and so both criminals will end up confessing and doing 10 years each, when they could have gotten away with just a year if they had both clammed up.

1471

The publication date of *Treatise on Arms*, by Diego de Valera, the first fencing manual. Fencing has its roots in ancient and medieval sword fighting, but it only **developed as a sport after firearms made swords obsolete**

3%

Boost to points total of an NFL team across a season, if they used game theory

10

Additional points over the course of a season an NFL team could achieve if they went from passing 56% of the time (the current average) to 70%

For a sport like fencing, game theory can suggest the strategy that optimizes a competitor's possible outcomes, by weighing up the risk:reward ratios of various strategies. For instance, it might turn out that the optimal strategy in the long-term is a counter intuitive one, such as exclusively employing high-risk lunges.

0.5

Increase in victories per season an NFL team could achieve if they adopted a game-theory approach to their passing game

2

Number of extra victories a year a MLB franchise could achieve by adopting a game-theory approach to pitching tactics

15

Number of runs a MLB team could avoid conceding if its pitchers threw 10% fewer fast balls—that's about 2% of their total runs conceded

What are the Odds? Calculating Odds

£1 billion

Amount bet on 2010 FIFA World Cup in UK alone (about US$1.5 billion)

US$330 billion

Value of global gambling, mostly internet and sports gambling

Odds are a way of expressing probability, or the likelihood of an outcome. They are important in sports as a way of estimating and appreciating certain outcomes, such as winning a race or achieving a certain score. Primarily they are important for sports betting, which accounts for the majority of the US$200 billion+ global gambling industry in 2012, according to German research firm Sport & Markt.

Odds mean different things depending on how they are presented. A horse racing bookie will typically give odds on a horse to win as, for instance, three to one. What this means is that the bookie estimates that the horse would win one out of every four races it enters. Expressed differently, he might describe the horse as having a 25% chance of winning. In this case the odds describe the ratio of favored outcomes (winning the race) to unfavored outcomes (losing the race).

However, this is not how odds are typically used in everyday conversation. If you were describing the chances of it raining tomorrow you might say there will be a "one in four" chance. This actually means the same as the bookie's "three to one," since you are claiming there will be a 25% chance of rain tomorrow. In this case the odds describe the ratio of favored outcomes to all possible outcomes.

Sports betting has become extremely complicated. It has evolved from direct booking with bookies who lay odds for bettors, to bettors making markets and laying odds for one another on betting exchanges. There are also many complex ways to bet, such as accumulator bets (where you bet on a series of outcomes and the odds multiply) and spread betting (where you win or lose more the further away the actual outcome is from the offered spread).

£2.8 billion

Value of sports-related gambling in England in 2008 (about US$4.2 billion)

1:6,000,000

Chances of being crushed to death by a vending machine while trying to get something out of it

1:280,000

Chances of being struck by lightning in the US in any given year

1:70,000

Lifetime odds of being killed by an asteroid impact

1:734,400,000

Annual risk of being killed by hailstones

1:11,000,000

Annual risk of being killed in a plane crash for the average American

20,000,000:1

Highest odds ever offered by bookmaker William Hill, on a novelty bet for Elvis Presley to ride into town on the Irish racehorse Shergar and play Lord Lucan in the Wimbledon tennis final

1:1.14

Chance that an American sports fan will eat at least one hot dog at a game in a year

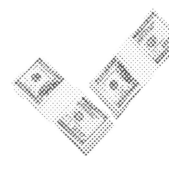

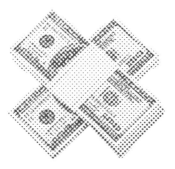

US$87.5 million

Amount legally wagered on the Super Bowl in Nevada in 2011

US$8 billion

Approx amount illegally wagered on the Super Bowl nationwide in 2011

US$1 million

Money bet on San Francisco 49ers to win the 1989 Super Bowl, by Las Vegas casino owner Bob Stupak—supposedly the single largest bet ever made in Vegas on anything

US$2.7 billion

Money bet with Nevada sports books in 2011

US$1.19 billion

Money bet with Nevada sports books in 2011 on college and pro football (44% of total)

WELCOME TO Fabulous LAS VEGAS NEVADA

1:7.69

Chance a baseball fan will perform some form of ritual to bring good luck to their team

#1

1:11,437

Odds that a boy in the US will grow up to be a major league baseball player

1:1.85

Odds that a MLB team will win its opening home game

No One Expects: Statistics, Upsets and the Most Interesting Sports

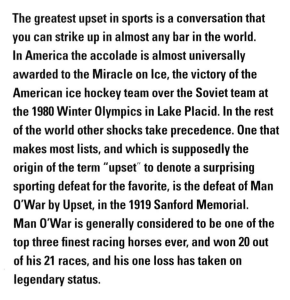

100:1

Odds against Upset when he took on Man O'War at the 1919 Sanford Memorial

91:1

Odds against Donerail, longest shot ever to win the Kentucky Derby, in 1913

The greatest upset in sports is a conversation that you can strike up in almost any bar in the world. In America the accolade is almost universally awarded to the Miracle on Ice, the victory of the American ice hockey team over the Soviet team at the 1980 Winter Olympics in Lake Placid. In the rest of the world other shocks take precedence. One that makes most lists, and which is supposedly the origin of the term "upset" to denote a surprising sporting defeat for the favorite, is the defeat of Man O'War by Upset, in the 1919 Sanford Memorial. Man O'War is generally considered to be one of the top three finest racing horses ever, and won 20 out of his 21 races, and his one loss has taken on legendary status.

Personal opinions are all very well, but what can mathematics tell us about upsets and surprises? The simplest mathematical test of an upset is to look at the odds against it. The longer the pre-event odds against the eventual winner, the bigger the upset. The topic is not purely academic. Being able to predict which sports and leagues are unpredictable could be a good measure of how enjoyable they are. Sports or leagues where the favorites always win are likely to be dull. Ones where there are regular upsets and against-the-odds victories should be more exciting.

A 2006 study by a team at Los Alamos National Laboratory in New Mexico analyzed results from more than 300,000 games over the last century from the U.S.A.'s national hockey, football, baseball and basketball leagues and the English Premier league. They found that the highest "upset frequency" was in the English Premier League, although their data indicated that in recent years MLB had overtaken the EPL. Perhaps English soccer is getting too predictable?

Goran Ivanisevic's world ranking just before he won Wimbledon in 2001, having needed a wild card just to get into the tournament

42:1

Odds against Buster Douglas before his 1990 world heavyweight boxing bout with Mike Tyson. Douglas knocked out Tyson in the 10th round

150:1

Odds against Joe Johnston winning the tournament before his triumph at the 1986 World Snooker Championship

400:1

Odds against Ireland at one stage during their match against England at the one day international cricket World Cup in Bangalore in March, 2011, before fighting back to win with five balls to spare

125th

45–2

Combined scores from seven previous international games/tournaments of the US team that went on to beat an English team packed with superstars by 1–0, in the greatest upset in FIFA World Cup history

8

Gold medals the Russian ice hockey team had won in the previous nine Olympics, before losing to the Americans in the Miracle on Ice, February 22, 1980

Playing at Home:
The Statistics of Home Advantage

1:1.82

Chances that an MLB team will win at home

Home advantage is regarded as a truism in sports. Fans, players and coaches assume it is real, bookmakers factor it into their odds and many tournaments and leagues offer home advantage as a reward for qualifying ahead of others. But assumptions about the efficacy and power of home advantage raise two questions: is it really happening, and if so why?

Statisticians have tried to answer these questions. On the first, there is strong evidence that home advantage is real, in that in sports, from baseball to soccer, teams win significantly more games at home than away. For a while even running events seemed not to be immune to the effect, with the vast majority of world records before WWII set by athletes competing in their native country.

The tougher question seems to be why playing at home constitutes an advantage. Analyses of Major League Baseball and the Bundesliga (the top soccer league in Germany) suggest the answer is not what people think.

Most people guess that the strength of home crowd support offers a psychological boost to the home team, but in fact there is no evidence to back this up (at least, not as a direct cause).

For instance, home advantage does not correlate with attendance. Other theories include familiarity with the home field and conditions, or relief from the rigors of traveling and being away from home. But none of these stand up to statistical analysis. Instead it seems that home advantage stems from unconscious umpire bias toward the home team. Home teams tend to get a favorable balance of marginal decisions. In baseball, for instance, the home team is more likely to get away with stealing a base and away batters are more likely to have strikes called against them. In football, home teams benefit from more time added on if they are behind and less if they are ahead, and they are more likely to get favorable penalty decisions.

80%

Proportion of world records set by athletes running in their native country before WWII

>25%

Proportion of world records set by athletes running in their native country today, probably because travel is easier and athletes are more accustomed to traveling

1:1.88

Chances that an American league team will win at home

1:1.76

Chances that a National league team will win at home

Intriguingly there may be a correlation between crowd fervor and the unconscious bias of referees. In the Bundesliga, the bias effect went up in stadiums without running tracks around the field, where the crowd is closer to the action and the referee. Referees were even more likely to make unconscious judgments about whether the game "deserved" to be played longer, with zero–zero draws being awarded less added-on time than draws with points on the scoreboard.

1 min

Estimated home bias amounts of addltional overtime when the home team is one goal behind, in stadiums in which the crowd is physically closer to the referee

BIAS IN THE GERMAN BUNDESLIGA

- 113 s: average bias in amount of time added-on when the home team is 1 goal behind rather than 1 goal ahead in Spanish soccer leagues
- 21 s: average bias in amount of time added-on in the German Bundesliga when the home team is 1 goal behind rather than 1 goal ahead
- 10 s: additional bias introduced in the last 5 rounds of the season compared to earlier rounds, possibly reflecting the "end-of-season effect," where crowds are more invested in the outcome
- 10 s: average time difference between scoreless draws and draws with scores. Matches with more shots on goal, more tackles and more crosses also last longer.
- x2: number of penalty shots awarded to the home team compared to the away team

Best of the Best: How Do You Measure Sporting Perfection?

7

Number of perfect 10 scores achieved by Nadia Comaneci at the 1976 Montreal Olympics

10

What constitutes perfection in sports? What is the perfect game, the perfect innings, the perfect race? As far as it is possible to answer, the response will be different for each sport. At the top of the list is baseball, in which it is actually possible to pitch a perfect game. This is when the pitcher strikes out or retires all of the opposition batters without giving up a single hit, or even letting a player get on base. Given that each team has nine innings with three outs in each, this means the pitcher has to retire 27 consecutive batters. The most recent occurrence was on June 13, 2012, when Matt Cain of the San Francisco Giants retired 27 consecutive Houston Astros batters, striking out 14 of them.

Can any other sport offer anything to match this? Events such as races, field events or soccer games pose problems. You can run faster than anyone else, throw further or score more goal/points, but theoretically you could have run faster, thrown further or scored more. One clue is that the perfect pitching game is an achievement

of negatives. The perfect pitcher allows no hits or bases. Parallels in other sports could come in something like tennis—if one player does not allow the other to score a single point in the entire match. A bowler in cricket is similar to a pitcher, and a bowler who bowled out the entire batting team without conceding a run would indeed have tasted perfection; however, this has never been done. The closest cricket can offer is the hat trick, where the bowler takes three wickets with consecutive balls. Hat tricks in soccer—in which a player scores three goals in a single game—are more common.

Another possible definition takes the broader perspective of a player's achievements in a year or over a whole career. Again, the achievements in question involve negatives—never having lost a boxing bout, not losing a single one of the major competitions in a year in tennis or golf, or not losing a match in a championship, as in European rugby's Six Nations when a Grand Slam is won.

9 9 9 9 9 9 9 9 9

9/9

Perfect scores for artistic impression awarded to Jayne Torvill and Christopher Dean at the 1984 Winter Games in Sarajevo

17

Games won out of 17 by the Miami Dolphins in 1972, the only football team ever to have a perfect season

22

Perfect games rolled in the Professional Bowlers Association

49

Wins out of 49 by Rocky Marciano, the only heavyweight boxer to have retired undefeated

3

Tennis players in the Open Era to win a Grand Slam, winning all four majors in a season: Rod Laver (1969), Margaret Court (1970), Steffi Graf (1988)

One Percent at a Time: Swimming and Performance Increments

0.27 s

Time difference between the gold and bronze medals in the women's 100 m freestyle swimming race at the 2008 Beijing Olympics

0.5%

Difference between gold and bronze medal winning performance in this race

Over the last two Olympic Games the most consistently impressive team in any sport for any nation has arguably been the British track cycling team. At the 2012 London Olympics, for instance, the British team won 7 out of 10 track cycling golds, and could have won more if not for some unfortunate disqualifications. What lay behind this awesome domination was a well-funded, micro-managed team structure, headed by performance director Dave Brailsford. Summing up the secret of his team's success, Brailsford told the BBC:

"The whole principle came from the idea that if you broke down everything you could think of that goes into riding a bike, and then improved it by 1%, you will get a significant increase when you put them all together. There's fitness and conditioning, of course, but there are other things that might seem on the periphery, like sleeping in the right position, having the same pillow when you are away and training in different places. Do you really know how to clean your hands? Without leaving the bits between your fingers? If you do things like that properly, you will get ill a little bit less. They're tiny things but if you clump them together it makes a big difference."

Making improvements 1% at a time doesn't sound like much, but it is clearly effective for the cyclists. In fact it would be effective for almost all the sports at the Olympics, because the margins between coming in first and finishing off the podium in fourth, let alone in second, are minute. Given that finishing on the podium is the only real measure of success for most elite Olympians, fourth place might as well be last for many. Yet in many cases the performance of the athlete who comes in fourth is just a tiny fraction less than the gold medallist. The table on the next page shows how a 1% performance would have affected the medal results of athletes in six events at the 2008 Beijing Olympics.

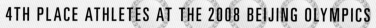

4TH PLACE ATHLETES AT THE 2008 BEIJING OLYMPICS

Event	Original time	1% faster	Result
Cycling: women's road race	3:32:28	3:30:52	Gold
Athletics: men's 100 m	9.93	9.83	Silver
Athletics: women's 400 m	50.01	49.51	Gold
Athletics: men's marathon	2:10:21	2:08:7	Silver
Swimming: women's 400 m	4:03:60	4:02:61	Gold
Kayak: men's K1 1,000 m	3:29.193	3:27.1	Silver

12%

Approx proportion
of athletes competing
who got on the podium

Approx proportion of
the 10,960 athletes
competing at the 2012
Olympics who won gold

88%

Approx proportion of
athletes competing who
went home empty-handed

4%

Athletes Uncovered: Unusual Olympic Statistics

The Olympic motto *Citius*, *Altius*, *Fortius* ("Faster, Higher, Stronger") could equally be Busier, Greedier, More Popular. Not only does each Games unleash a torrent of stats on the track and field, velodrome and swimming pool, rowing lake, boxing ring and many other venues, the Olympics rack up impressive numbers in a host of other ways.

Typically Olympic Games are sited in an urban district in need of regeneration, with the billions of dollars spent on infrastructure and venues intended to leave a lasting physical legacy. At the 2012 Olympics, for instance, an entire derelict post-industrial neighborhood of London was regenerated alongside construction of at least four major venues. Everything else about an Olympic Games is also on an Olympian scale. The London games saw the world's biggest McDonald's operating at the Olympic Park, where there were 10,000 toilets installed. Over a billion people tune in for opening ceremonies, and venues can expect hundreds of thousands of visitors in total.

A crucial focus here is that the number of athletes attending each Games tends to increase. The London Olympics saw the largest contingent of athletes yet, with almost 11,000 of them. Not surprisingly, feeding thousands of elite athletes calls for epic quantities of food. The "Food for Sports" numbers on the next page show just how large a intake was required to energize contesting top-class athletes.

But which nation dominates the Olympics? The statistics on pages 168–169 look at Olympic teams and athletes, showing that the biggest countries, the Olympic superpowers of the US, UK, Russia, China, Germany and Australia, typically account for one in four athletes. However, when team sizes are compared to the size of the countries' populations, an interesting anomaly is revealed: the Cook Islands have the highest density of Olympic athletes of any nation in the world!

FOOD FOR SPORTS

45,000

Number of meals athletes
consumed per day

23 tons

Amount of cheese
per day

90 tons

Amount of seafood per day

110 tons

Meat per day

331 tons

Fruit and vegetables
per day

25,000

Loaves of bread
per day

OLYMPIC TEAMS AND ATHLETES

40/100,000

Representation rate of athletes per population for the Cook Islands at 2012 Olympics

0.889/100,000

Representation rate of athletes for the UK, highest of the major countries

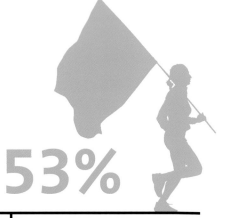

44%

Proportion of female athletes for Olympics as a whole at 2012 Games

25%

Olympic athletes typically coming from the big six teams: US, UK, Russia, China, Germany and Australia

53%

Proportion of females in Japanese team, the team with the highest proportion of female athletes

OLYMPIC SUPERLATIVES

22

Total number of medals won by Michael Phelps, including 18 gold, making him the most decorated Olympian ever

8

Gold medals won by Michael Phelps at the 2008 Olympic Games, the most in a single Games

776 B.C.E.

First recorded Olympic Games

1920

Year of the last Games that included tug-of-war as a sport

1948

Year of last Games that included an artistic competition, with medals for painting, sculpture and architecture

Animal Olympics: Athletic Records in the Animal and Human Worlds

Humans are extraordinary and probably unique in the range and versatility of their sporting prowess, capable of running, swimming, throwing, skiing, shooting, lifting, balancing and all the other myriad skills and abilities they demonstrate. But the bare fact is that their stats pale by comparison with the records stacked up by other species.

MAXIMUM SPEED

23.4 mph
Human

64 mph
Cheetah

55 mph
Racehorse

67 mph
Sailfish

5.8 s
Time for cheetah to run 100 m

9.58 s
Usain Bolt's world record for 100 m

JUMPING

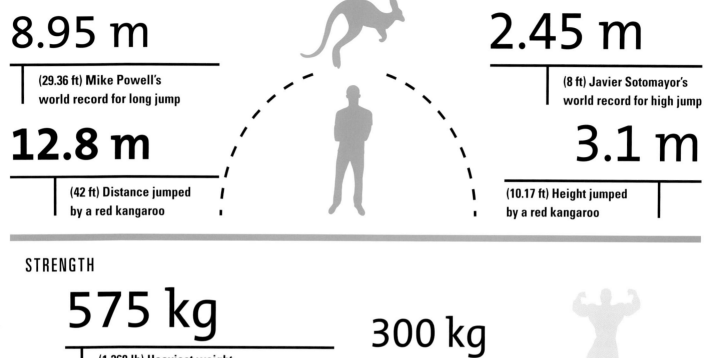

8.95 m
(29.36 ft) Mike Powell's world record for long jump

12.8 m
(42 ft) Distance jumped by a red kangaroo

2.45 m
(8 ft) Javier Sotomayor's world record for high jump

3.1 m
(10.17 ft) Height jumped by a red kangaroo

STRENGTH

575 kg
(1,268 lb) Heaviest weight lifted by a human

300 kg
(661 lb) Weight an African elephant can lift with its trunk

820 kg
(1,808 lb) Weight an African elephant can carry

900 kg
(1,984 lb) Weight a gorilla can lift

Money Makes the World Go Round: Sporting Salaries

€3.2m

(About US$4.16 million) Average salary at Kolkata Knight Riders, Indian Premier League T20 cricket franchise, ranking 36th in world of highest non-soccer, -baseball or -American football teams

€4.7m

(About US$6.17 million) Average salary at LA Lakers (NBA) and NY Yankees (MLB), the highest-paying sports teams outside of soccer

The number that talks loudest in the modern world of sports is the one on the paycheck. Professional athletes are paid astronomical sums of money. The top ten earning English Premier League players now all make more than a FTSE100 chief executive, and the average EPL—salary £1.162 million/year (about US$1.7 million) dwarfs the average annual UK salary of £26,200 (about US$39,300). The average MLB salary is US$3.44 million, compared to the median annual US salary of US$26,364.

How did the financial world of sports players become so disconnected from that of the ordinary citizen? First, it is important to note that it was not always so. Data presented on page 173 show how weekly salaries in the Premier English football league, which offers an illuminating example, increased exponentially over the last century or so. Up until 1961 there was a wage cap, and before this wage limits for players were in line with working men's pay in other professions, with a very slow rate of increase. It took 46 years for the weekly wage limit to go from £4/wk to £12/wk.

Without the wage cap salaries started to go up faster, and accelerated further in the 1980s as the influence of big money in the Italian Serie A was felt, driving up weekly wages above the five-figure mark. The third great shift occurred in 1995 when a landmark court ruling known as the Bosman ruling gave players greater rights to negotiate big salaries, and the highest weekly wage rocketed from £10,000 to £40,000 in the space of two years. The open nature of the British economy has made it easy for foreign billionaires to buy clubs, and they have unleashed yet more torrents of money, driving weekly wages above the £200,000 mark (about US$300,000).

WAGE INFLATION IN ENGLISH SOCCER

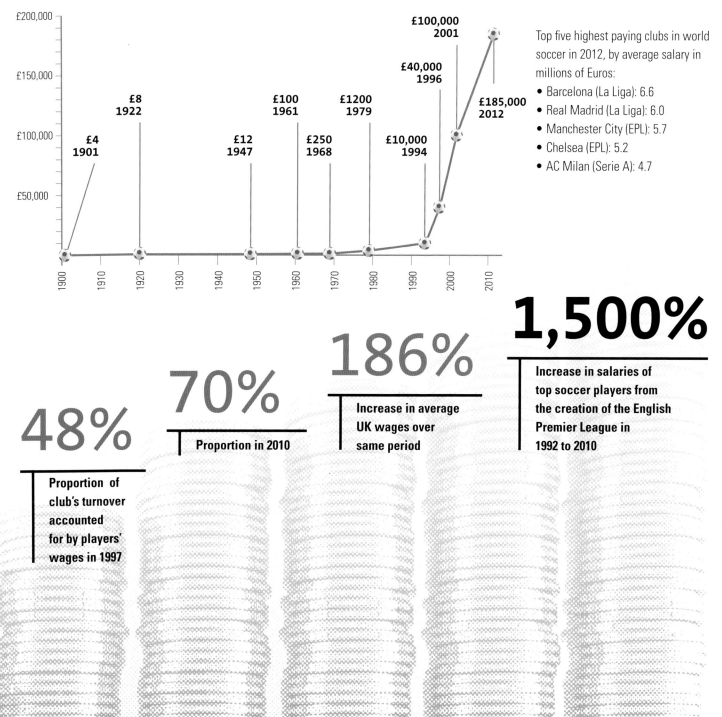

£200,000
£150,000
£100,000
£50,000

£4 1901
£8 1922
£12 1947
£100 1961
£250 1968
£1200 1979
£10,000 1994
£40,000 1996
£100,000 2001
£185,000 2012

1900 1910 1920 1930 1940 1950 1960 1970 1980 1990 2000 2010

Top five highest paying clubs in world soccer in 2012, by average salary in millions of Euros:
- Barcelona (La Liga): 6.6
- Real Madrid (La Liga): 6.0
- Manchester City (EPL): 5.7
- Chelsea (EPL): 5.2
- AC Milan (Serie A): 4.7

48%
Proportion of club's turnover accounted for by players' wages in 1997

70%
Proportion in 2010

186%
Increase in average UK wages over same period

1,500%
Increase in salaries of top soccer players from the creation of the English Premier League in 1992 to 2010

US$42.5 million

Cristiano Ronaldo, Soccer

US$42.4 million

Peyton Manning, Football

US$46 million

David Beckham, Soccer

The massive disparity between sporting salaries and average salaries causes many negative reactions, with resentment, fears about unsustainable debt, and qualms about the spiritual and civic damage inflicted. Can the sums involved possibly be justified? The EPL argues its position by pointing to continued high levels of attendance (consistently around 90% for the last 15 seasons) and growth of TV audiences, and with sports teams increasingly being developed as brands in both traditional home markets and lucrative new markets in Asia, there is supposedly an economic justification for spending big on elite players. Players have also pointed to the brief extent of even an injury-free career, not to mention the uncertainty of a career-ending injury. Ultimately players and agents can argue that they charge only what the market will bear, and teams that pay too much and have an unsustainable business model will fail. But as long as team profits continue to increase, wage inflation will almost certainly continue.

US$53 million

LeBron James, Basketball

US$52.3 million

Kobe Bryant, Basketball

US$52.7 million
Roger Federer, Tennis

US$47.8 million
Phil Mickelson, Golf

US$59.4 million
Tiger Woods, Golf

US$62 million
Manny Pacquiao, Boxing

US$85 million
Floyd Mayweather, Boxing

£1.32 million
(Approx US$2 million) Amount per punch earned by Audley Harrison in his fight with David Haye in November 2010, in which Harrison landed just one punch (and lost), making him the highest-earning boxer of all time by punch

175

A Hard Rain: Duckworth-Lewis

An impressive example of the application of mathematics to sports comes from the world of cricket. In the long form of the game, known as test cricket, a draw is a possible result and if the match is interrupted or even washed out by bad weather a draw is the usual result. In the limited overs form of the game, however, a draw is not an option, as the game must always end with a result (although it can end with a tie, which is not the same as a draw. A tie results when both teams end up with exactly the same amount of runs, which is very unusual).

This posed a problem for the cricket authorities, as limited overs matches are frequently disrupted by rain, especially in England. (An "over" is six balls bowled consecutively. In a limited overs match, only a set number of overs are bowled to each side.) If the team batting second ends up with fewer overs in their innings owing to rain interruption, then to ensure a fair result they must be set a revised target to reach.

Statistician Frank Duckworth and mathematician Tony Lewis came up with a solution that took into account both overs remaining and wickets in hand—the two factors that make up what they consider to be a team's "resources." They developed tables showing what percentage of resources corresponded to what combination of overs and wickets left, and used these as the basis for a simple system: the Duckworth-Lewis Method (DLM).

In the DLM, the "resources" lost to rain are taken from the table, and subtracted from the affected team's total resources. Their revised target is then calculated by multiplying the original target by the ratio of the affected team's resources to the unaffected team's resources, which gives the score for a tie, and adding one for the winning score.

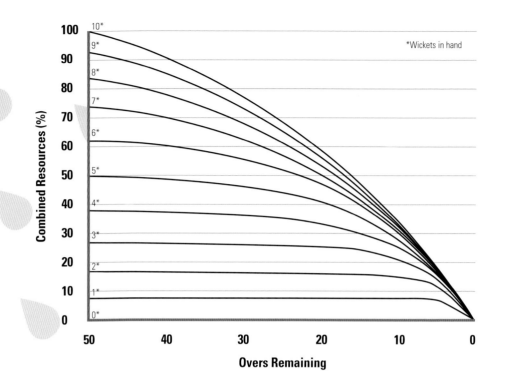

Combined Resources (%) — vertical axis
100, 90, 80, 70, 60, 50, 40, 30, 20, 10, 0

Curve labels (Wickets in hand): 10*, 9*, 8*, 7*, 6*, 5*, 4*, 3*, 2*, 1*, 0*

*Wickets in hand

Overs Remaining — horizontal axis
50, 40, 30, 20, 10, 0

Example

- Team A bats first and scores 276
- Team B reach 224 for 7 after 40 overs but then rain stops play, ending the match
- According to the table, having 3 wickets in hand and 10 overs left corresponds to 20.6% of resources
- Team B lost 20.6% of the resources they had to win the match, so the total resources available were 100 - 20.6 = 79.4%
- The ratio of resources of Team B:Team A = 79.4 / 100 = 0.794
- The revised target is thus calculated as (276 x 0.794) + 1 = 220
- Team B are ahead of this target and are duly declared the winners

RUN RATE METHOD

The original solution to the interrupted innings problem was to use a crude calculation of run rate to revise the target. If Team A had scored 250 runs in 50 overs at a rate of 5 runs per over, and Team B lost 25 overs to rain, their target would be reduced by 25 x 5 = 125 (i.e., they would have to make 125 to win). This produced unfair results. A team in the example scenario that reached 126 for 9 after 25 overs would be declared the winner, despite having no hope of getting the remaining runs with their last wicket partnership.

MODEL FOR CALCULATING RESOURCES TABLE

The model is the two factor exponential relationship: $Z(u,w) = Z_o F(w) [1 - \exp\{-bu/F(w)\}]$ where $Z(u,w)$ is the average further number of runs expected to be made when there are u overs remaining and w wickets down. $Z_oF(w)$ is the asymptotic value of further runs expected with w wickets down as u tends to infinity, F(0) being set to unity. The parameters, b, Z_o and the nine values of F(w) were estimated from an analysis of a one-day database.

Spotting a Winner: Tour de France

250

Approx number of
riders who have
worn the Yellow
Jersey at some
stage in a Tour

The Tour de France is the biggest annual sporting event in the world. To understand the scale and mystique of the race you need to look at some of the incredible numbers that surround it, from its dizzying length to the awesome calorie consumption of the contenders, from the army of people following the tour to the handful of people who have died watching it (more than have died taking part!).

One of the most important numbers is four—the number of main jerseys up for contention. The simplest one to explain is the white jersey, awarded to the leading young rider (under 25) in the race. The yellow jersey is worn in each stage by the current leader of the general (overall) classification (GC). The GC is decided by cumulative time over all the stages so far, although an important point to remember is that every rider in the peloton, or main group, is awarded the same time as the leading rider of the peloton. Since there may be dozens of riders in the peloton it is impractical to separate them and would cause a dangerous scramble as they cross the finish line.

The green jersey, aka the points jersey, is won by the rider who accumulates the most points, which are awarded for position at intermediate checkpoints and the ends of stages. It is contested by sprinters. These are riders built for power over short bursts, which in turn makes them uncompetitive in the tough mountain stages, where lighter riders have the advantage (see page 32 for an explanation why).

The green jersey contenders are therefore unlikely to ever be in contention for the yellow jersey, because in the mountain stages they struggle to keep up with the rest of the Tour and have to work hard just to avoid the dreaded *voiture balai*, or "broom wagon," which "sweeps up" straggling riders who fall too far behind, resulting in their disqualification. Green jersey contenders lose so much time on mountain stages that they are usually way off the pace in the GC.

2,173 mi

Total distance of 2012 Tour—equivalent to distance from London to Cairo or Tel Aviv

1,434 mi

First and shortest Tour length

3,570 mi

Longest Tour in 1926

16

Most Tours ridden by one rider, by Joop Zoetemelk, between 1970 and 1986. He finished every one

4

Contenders who have died during the Tour de France. Twenty-six bystanders, officials and other non-contenders have also died during the Tour

10,000

Total number of riders to compete in the Tour

34

Most stage wins, by Eddy Merckx

20

Age of youngest Tour winner, Henri Cornet in 1904

8 s

Shortest winning margin, between Greg LeMond and Laurent Fignon in 1989

36

Age of oldest Tour winner, Firmin Lambot (36 yrs) in 1922

The polka-dot jersey is awarded to the king of the mountains, the rider who wins most points in the mountain stages. Points are awarded to riders according to their position as they reach the top of the different climbs along the stage (i.e., most points for first, less for second, etc.), with double points for a stage finish that happens on a climb. The tougher the classification of the climb, the more points are on offer for those racing up it. Polka-dot jersey contenders have to be light and built for endurance, which means they are not competitive on the sprint stages and time-trial stages. They generally lose so much time on time-trial stages that they cannot be in contention in the GC and thus cannot win the yellow jersey.

There are other jerseys. The reigning World Road Racing Champion is entitled to wear a rainbow jersey, while national champions can wear their nations' racing colors. The rider rated the most "combative"—i.e., who has been bravest and boldest in the previous stage—wears white race numbers on a red background instead of the standard black on white. He used to be given a different-colored jersey, but race organizers found that this was like putting a target on his back as it made him a focus for attacks by other riders.

Typical number of chains used by a single rider during the race

3

118,000

Total calories burned by a typical Tour finisher

20%

Energy saved by riding in teammate's slipstream

36

French wins in Tour

5,900

Calories per day burned by a Tour rider on average

2 billion

Global audience for Tour on 78 television channels in 190 countries

30–40 s

Time for peloton to pass a given point

15 million

Spectators lining the route

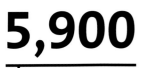

Marathon Men: Soccer Players' Exertion

18 kCal/min

Energy
consumption rate

1:2

Ratio of high exertion: rest, i.e., for every
10 seconds spent jogging/sprinting, a player
gets an average of 20 seconds rest

Anyone who watches the exploits of long-distance runners circling the track for 10,000 m (6.2 mi) or running a 42.2 km (26 mi, 385 yd) marathon has marveled at the extraordinary fitness that enables an athlete to still have the energy to sprint for the line after running farther than the vast majority of the population could manage. Similarly, competitors in other endurance sports like the Tour de France, the triathlon or the open water swim impress us with their fitness. Fewer people, perhaps, appreciate the astonishing fitness of professional soccer players.

Thanks to advances in monitoring technology, it is now possible to analyze to a greater extent than ever before the exertions of players during a game, to break down their movement and activities and examine them in minute detail. Combining camera tracking technology with computer algorithms that model the field environment and translate recorded motion into distance and speed, with GPS tracking also increasingly being used, it is now possible to work out exactly how much time soccer players spend doing different activities, how far and fast they run and how much energy they expend.

Soccer players might seem to spend a lot of time standing around or moving at a walking or jogging pace, and the numbers reveal that this is true: two-thirds of the match is spent walking or jogging. But the numbers also show that players cover between 6 and 9 mi in a match, most of it at a sprint. They perform hundreds of sprints and turns. But perhaps the most surprising thing the numbers reveal is that for all their huffing and puffing, players spend a vanishingly small amount of time with the ball. In the entire match they might get just a few seconds worth of ball time with which to influence the outcome, and this is where fitness is so important. Soccer players need to be able to sprint at speeds of over 12 mph hundreds of times, covering thousands of yards in total, and still be able to focus and perform for those crucial moments that might not come until the end of the game.

80–110

Sprints per game

3,900

Jumps per season

90–140

Interactions with
ball per game

550

Turns per game

60–90 s

Total time spent with ball
over the course of 90 min

Distance covered (mi)		Fast running distance (mi)	
	6.2–9.3		1.5
	3.7–4.3		0.9

Time spent standing/walking (min)		Sprint distance (ft)	
	59 (65%)		2,133
	64 (75%)		1,027

Time spent jogging (min)		Work to rest ratio	
	25 (27%)		1:2
	15 (19%)		1:3

Time spent sprinting (min)		Energy burned (kCal)	
	7:40 (8%)		1,400
	4:45 (5.7%)		1,900

■ Soccer player
▨ Rugby forward

Mathematics, Biology and the Environment

4

SPORTS DO NOT EXIST IN A VACUUM. ATHLETES COMPETE WITHIN SPECIFIC ENVIRONMENTAL CONSTRAINTS AND ARE LIMITED BY THE BASIC FACTS OF THEIR BIOLOGY, AND THESE CONSTRAINTS AND LIMITATIONS CAN BE QUANTIFIED. THIS SECTION TAKES A NUMERICAL LOOK AT CONSIDERATIONS OF NATURE AND PHYSIOLOGY, FROM THE OCEAN DEPTHS TO THE TOP OF THE ATMOSPHERE.

A Man for All Surfaces: Effects of Tennis Court Surfaces

In the history of tennis, a career grand slam of major titles has been extremely rare. In recent years only Andre Agassi, Roger Federer, Rafael Nadal and Serena Williams have managed it. Tennis greats like Pete Sampras, John McEnroe and Björn Borg never managed it. The reason the feat is so rare is that the majors are played on different surfaces. Wimbledon is played on grass, the French Open on clay and the US and Australian Opens on synthetic hard courts. Each of these surfaces has different properties, different speeds and favors different playing tactics and player capabilities.

The International Tennis Federation classifies the different court types according to their court pace rating (CPR), a measure of surface friction and bounciness that determines whether the ball skids on and stays low, or sits up and bounces high, factors which in turn determine whether the court is perceived as playing fast or slow. Technically speaking, the CPR is calculated according to an equation relating the coefficient of friction of the surface (how "sticky" it is and therefore the degree of horizontal velocity an incident ball will retain after bouncing) to the coefficient of restitution (COR see page 46).

Grass courts have the lowest COR because the grass stalks cushion the arriving tennis ball, depleting the amount of downward velocity that can be converted into upward velocity on rebound. The grass also retains moisture on its surface, reducing friction so that the ball retains more of its horizontal velocity. The combination makes grass play faster, because the ball flies on and stays low, giving players less time to make their shots. Clay, on the other hand, has high friction and a high COR, making a player's power less important since the surface will tend to take the sting out of their shots, but making spin and court positioning and movement more important, over longer rallies. The properties of clay particularly favor the playing style and attributes of Rafael Nadal, who produces very high spin rates, but it is a testament to his talent and power that he has also triumphed on the other two surfaces.

18

Number of games you can use a tennis ball on grass before its aerodynamic properties change due to wear

23

CPR for clay

CAREER SLAM WINNERS

Male	Female
Andre Agassi; Don Budge; Rafael Nadal; Roy Emerson; Rod Laver; Roger Federer; Fred Perry	Serena Williams; Maureen Connolly; Margaret Court; Steffi Graf; Martina Navratilova; Doris Hart; Shirley Fry; Billie Jean King; Chris Evert

7

Number of male tennis players who have won career grand slams

9

Number of female tennis players who have won career grand slams

2,700 rpm

Spin rate of average professional men's tennis player forehand

COURT PACE RATING CATEGORIES

Slow	≤29
Medium-Slow	30–34
Medium	35–39
Medium-Fast	40–44
Fast	≥45

35

CPR for hard courts

46

CPR for grass

SPORTS BY NUMBERS
Tennis

COURT

The court shall be a rectangle, 78 ft (23.77 m) long and, for singles matches 27 ft (8.23 m) wide. For doubles matches, the court shall be 36 ft (10.97 m) wide

LINES

The center service line and center mark shall be 2 in (5 cm) wide. The other lines of the court shall be between 1 in (2.5 cm) and 2 in (5 cm) wide, except that the baselines may be up to 4 in (10 cm) wide

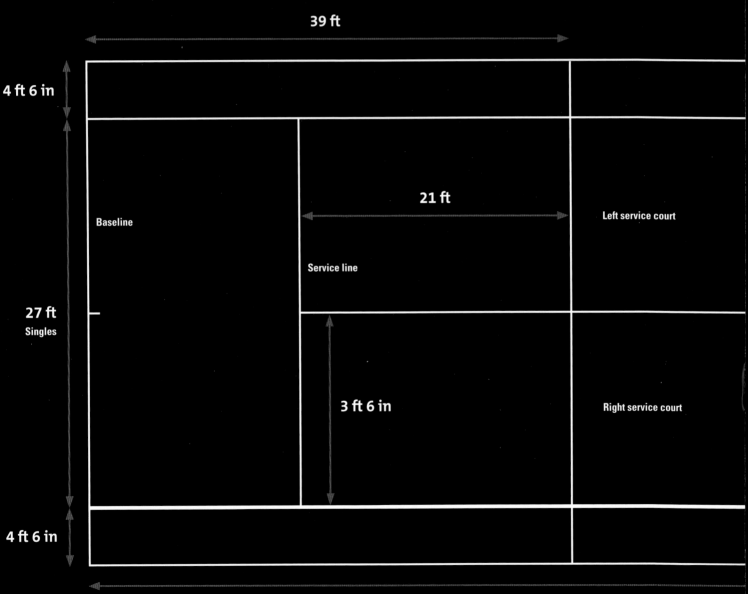

39 ft

4 ft 6 in

21 ft

Left service court

Baseline

Service line

27 ft
Singles

3 ft 6 in

Right service court

4 ft 6 in

78 ft

NET

Two net posts at a height of 3½ ft (1.07 m). The height of the net shall be 3 ft (0.914 m) at the centers. The maximum diameter of the cord or metal cable shall be ¾ in (0.8 cm). The centers of the net posts shall be 3 ft (0.914 m) outside the doubles court on each side

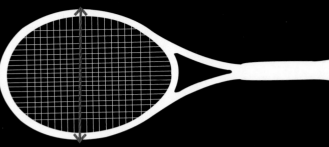

12.5 in

RACKET

The frame of the racket shall not exceed 29 in (73.7 cm) in overall length, including the handle. The frame of the racket shall not exceed 12.5 in (31.7 cm) in overall width. The hitting surface shall not exceed 15.5 in (39.4 cm) in overall length and 11.5 in (29.2 cm) in overall width

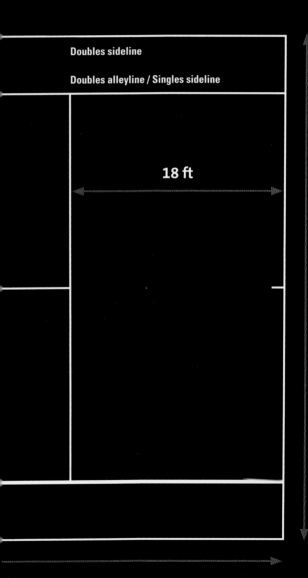

Doubles sideline

Doubles alleyline / Singles sideline

18 ft

36 ft

Doubles

BALLS [USED AT WIMBLEDON]

Tennis balls are rigorously inspected before they are allowed to be used in a match at Wimbledon. Each ball must bounce between 53 in and 58 in when dropped from a height of 100 in onto a concrete surface. Balls must be 2½ in in diameter and they must weigh 2 oz each

2½ in

50,000

Number of tennis balls used over the three weeks that Wimbledon runs every year

The Right Tools for the Job: Materials Science and Sports

100:1

Even "pure" sports such as running and swimming depend to a surprising degree on technology. Modern running shoes, for instance, are built from advanced materials and designed on scientific principles. Other sports are largely defined by the tools involved—rackets for tennis, clubs for golf, skis for skiing and bicycles for cycling. Such sports have advanced far beyond the limits of their origins thanks to technology, and in particular thanks to materials science.

Materials science is the study of materials, particularly their strength, toughness and stiffness. These might sound like different words for the same thing, but in materials science they have distinct and clearly defined meanings, with important implications for sports. Strength is the ability of a material to withstand applied force without failing—for instance, how hard it can be stretched or squeezed before breaking; the opposite of strong is weak. Toughness is the ability of a material to absorb energy without cracking; the opposite of tough is brittle. Stiffness is the ability of a material to withstand bending; the opposite of stiff is flexible.

Finding the right material for a piece of sports equipment involves considering all these factors, along with density (which will determine how heavy the material is) and cost. The graph on the right shows how different materials compare in terms of the relation between stiffness and density, and strength and toughness. Imagine designing a tennis racket; it needs to be strong and tough, stiff and light. You can use the graph to find the material that suits your needs best. The graph shows that wood is light while also being relatively stiff, but that composites (such as carbon fiber composites or CFC—commonly called graphite) offer extreme stiffness without being too heavy. The relative weakness of wood meant that wooden rackets had to be small and thick to resist breakage, and this meant they were heavy with small sweet spots. The introduction of composites made it possible to have larger, lighter rackets with bigger sweet spots than wooden rackets.

Ratio of stiffness (in gigapascals) to density (in tons per cubic meter) of CFC—making this the best material for tennis rackets

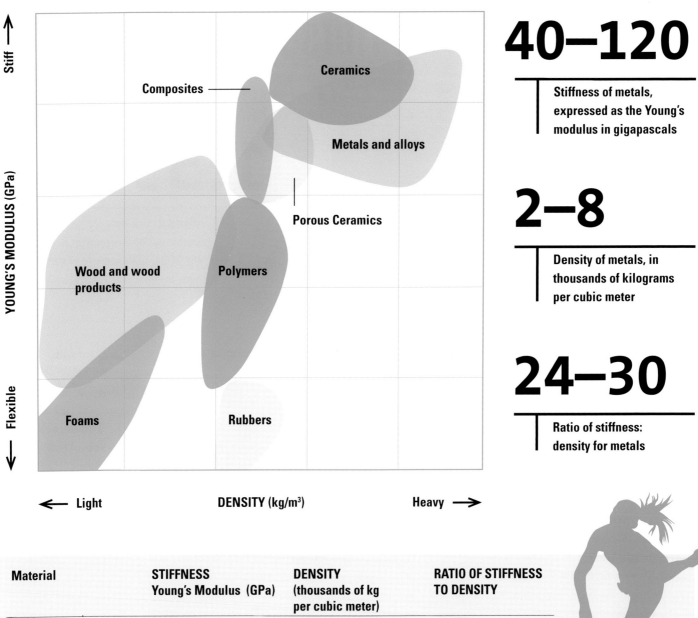

40–120

Stiffness of metals, expressed as the Young's modulus in gigapascals

2–8

Density of metals, in thousands of kilograms per cubic meter

24–30

Ratio of stiffness: density for metals

Material	STIFFNESS Young's Modulus (GPa)	DENSITY (thousands of kg per cubic meter)	RATIO OF STIFFNESS TO DENSITY
Metal	40–210	~2–8	24–30
Glass	73	~2.5	~30
Ceramic	400–700	~3.5	100–230
Carbon-Fiber Composite	200	2.0	100
Wood	14	0.5	28

Around the Oval: Horsepower

1:59.40

World record time for the Kentucky Derby, set by Secretariat in 1973, the first horse, and one of only two, ever to run under 2 mins

2:24

Secretariat's world record time in the Belmont Stakes in 1973

Flat racing horses are thoroughbreds, a breed that traces its ancestry back to three horses from Asia introduced to England in the 17th and 18th centuries, which were crossbred with native horses. Over three centuries of selective breeding have resulted in horses specialized for maintaining high speeds over long distances.

Horsepower is a measure of work done over time, a way of looking at power output. The term was originated by James Watt, the steam engine pioneer. His new engines were intended in part to replace working horses, so to provide a useful point of reference for marketing, Watt derived a unit of power output based on his observations of working horses. Using the imperial measurements standard at the time, Watt calculated that a single horse could do work equivalent to raising 33,000 lb (14, 969 kg) of weight a foot (30 cm) off the ground in a minute. In the coal mines that Watt may have been observing at the time, this equated to a single horse operating a turnstile or mill to raise a 100 lb (45 kg) bucket of coal 328 ft

(100 m) in a minute. Whether or not such output is sustainable by a horse is a topic of debate, but according to at least one study of 19th-century workhorses it may be accurate.

Racehorses generate high levels of power output, but what is most astonishing is their ability to sustain it over long periods. Quarter horses, the fastest breed over short distances, can reach over 50 mph (22.35 m/s), but the best pace most horses can expect to maintain over two or three furlongs (1 furlong = 201 m) is around 33 mph (16 m/s). This puts into perspective the feats of the legendary Secretariat, widely regarded as the fastest and best flat racing thoroughbred in history. Secretariat maintained an average speed of 33 mph (16 m/s) over 12 furlongs when he won the Belmont Stakes in 1973. This is the equine equivalent of Usain Bolt running the 800 m at his 100 m pace.

746 W

33,000 ft-lb/min

Standard unit equivalent of 1 hp—approx equivalent to the rate at which energy is supplied by sunlight falling on a square meter of turf at Churchill Downs, home of the Kentucky Derby

The work rate equivalent of 1 horsepower (hp)

0.1 hp

Output a healthy human can sustain

2.5 hp

Peak output sustainable by an athlete

JUST HOW FAR AHEAD WAS SECRETARIAT?
BELMONT STAKES WINNERS

Crusader
Record Holder: 1926
42 lengths behind

Ruler on Ice
Winner: 2011
34 lengths behind

Citation
Record Holder: 1948
22 lengths behind

Gallant Man
Record Holder: 1957
13 lengths behind

Secretariat
Record Holder: 1973

Slipstreaming: Cycling and Aerodynamics

33%

Energy saving of slipstreaming cyclist in the Tour de France

6.5%

Metabolic savings from running 1 m (39 in) behind another middle distance runner

Racing athletes don't just have to compete with one another, they also have to overcome air resistance. Air might seem insubstantial but it has mass—at sea level a cubic yard of air weighs 2 lb 2 oz. This mass slows down runners and riders in two ways.

First, through direct friction. This is where the air molecules physically bang into and rub against the athlete and his equipment. Athletes often wear skin-tight suits to minimize this element of air resistance. Secondly, drag. Drag is what happens when air cannot instantly get out of the way of the runner/rider. Instead it piles up in front, while behind there is relatively less air. This in turn leads to high pressure in front and low pressure behind, resulting in a force pushing/pulling the athlete backwards—hence, drag. Drag accounts for up to 90% of the air resistance affecting a cyclist.

One way to overcome the effects of drag is to run or ride in someone else's slipstream: the region of low pressure behind the leader. Slipstreaming athletes save energy—

at the speeds involved in cycling, energy savings can be 33% or more. Obviously the slipstreamer cannot go faster than the leading rider, but if the leader is a powered vehicle capable of great speed then the slipstreaming rider can go correspondingly fast.

7.5%

Proportion of total energy spent overcoming air resistance while running at 20 fps

13%

Proportion of energy spent overcoming air resistance while running at 33 fps

c. 1 in

Distance from leading bicycle needed to gain maximum benefit from slipstream

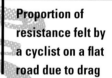

70%–90%

Proportion of resistance felt by a cyclist on a flat road due to drag

Bicycle length
Approximate extent of slipstream benefit from preceding rider

0.75
Drag coefficient for a racing cyclist

44.3 mph
World speed record for an unassisted bicycle

152 mph
World speed record for cyclist riding in slipstream of specially designed car

In at the Deep End: Mathematics of Swimming

1.5%

Efficiency of swimming at low speeds, in terms of ratio of energy consumed to swimming power generated

Humans are relatively slow swimmers. Even the fastest humans can manage only a fraction of the speeds achieved by many marine organisms. The fastest fish—the sailfish—is more than 14 times faster than the fastest human swimmer, and it can swim roughly as fast as a cheetah can run. Humans are also inefficient swimmers, with poor efficiency rates of less than 2% at slow speeds, some way off the 25% efficiency that bicyclists can achieve.

This is not surprising, as swimmers have to overcome several handicaps. The human body is not designed for fast swimming. The density of the body is roughly equal to freshwater and slightly less than seawater, so humans float at the surface, creating a bow wave as they swim, which slows them down. Also the lightest end of a human is the head end, which is helpful for keeping the nose out of the water but gives the feet a tendency to sink. This in turn increases the frontal area presented to the direction of travel, increasing the drag that slows the body. Women's legs tend to have more fat, making them more buoyant so they sink less, which in turn makes the average woman a more efficient swimmer than the average man.

The real problem with water, though, is its high viscosity. This means that hydrodynamic drag is a much bigger force than aerodynamic drag. Marine mammals overcome this problem with their streamlined shapes, adapted by evolution, while their fins and tails suit hydrodynamic propulsion much better than human hands and feet. Fins have broad areas to displace the maximum amount of water with each stroke, but present thin, streamlined cross-sections in the direction of travel, reducing drag. Sharks have yet another adaptation—their skin is made up of hard scales covered in tiny ridges, which create minute vortices in the water that dramatically reduce drag (perhaps akin to the seam on a swinging cricket ball—see page 110). Swimming suits modeled on sharkskin are supposed to boost the speed of human swimmers, although not all researchers are convinced they make a difference.

8%

Efficiency of swimming at high speeds

20.91 s

World record for men's 50 m freestyle, set by Cesar Cielo Filho in 2009, swimming at a speed of 2.39 m/s (7.84 fps)

23.73 s

Women's world record for 50 m freestyle, set by Britta Steffen in 2009, swimming at a speed of 2.1 m/s

880 J

Energy expended per meter (39 in) traveled per square meter (1.2 sq yd) of body surface for average male swimmer

630 J

Energy expended per meter (39 in) traveled per square meter (1.2 sq yd) of body surface for average female swimmer—women are more efficient swimmers

100 N

Force needed to tow a man through water at a sprinting speed of 2 m/s (6.6 fps)

11 N

Force needed to tow a man at 2 m/s (6.6 fps) if he were dolphin-shaped, owing mainly to streamlining

Get a Grip: Tires in Auto Sports

10,000 mi

Minimum life expectancy of an ordinary car tire

We've seen how Formula 1 (F1) cars can achieve remarkable cornering speeds far higher than those of other racing cars, let alone normal road cars, thanks to their ability to generate massive levels of grip. Grip is the degree to which the car sticks to the road and can resist sliding off. It has two main components: aerodynamic and mechanical grip.

Aerodynamic grip is the same as the lift force that keeps an aircraft in the air or makes a curveball bend (see page 18). F1 cars have wings and fins that generate very high levels of lift—more than enough for them to take off and fly if they were put on the wrong way up. But since F1 wings are fitted upside down compared to aircraft wings, the force they generate pushes the car down, helping it to stick to the road even at ferocious speeds. F1 cars generate so much aerodynamic grip that they could drive upside down, while the upward "suction" acting on the road surface beneath is so great that on the street circuit of Monaco the manhole covers would be sucked out of the road if they were not welded on before the race.

Mechanical grip is created by the friction between the surface of the tire and the road surface. F1 tires are technological marvels, precision-engineered to be as sticky as possible over as much road surface as possible, without slowing down the car too much. They are concocted from high-performance compounds of up to a hundred different substances, and designed to operate at high temperatures. The pressure inside an F1 tire must be minutely controlled, since the slightest deviation can ruin the tire's performance profile.

Even small changes to the formulation of a tire can have significant effects on its performance characteristics. The major road vehicle tire manufacturers, for instance, introduced new silicon-based compounds in the 1990s, which reduced the "rolling resistance" (friction acting against forward movement) of tires by up to 30%, making them more fuel-efficient and saving billions of gallons of fuel for drivers around the world.

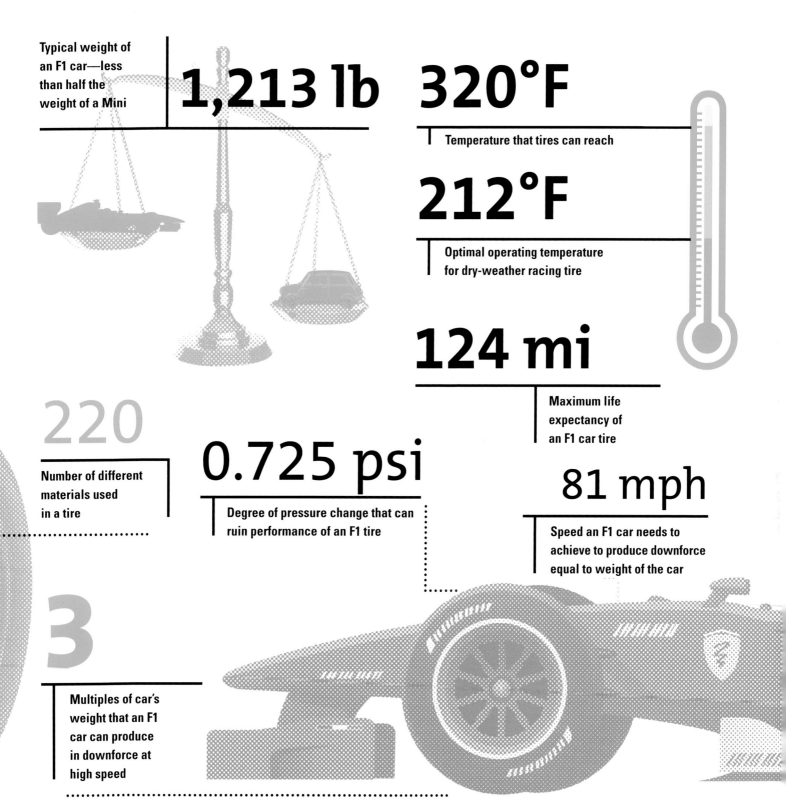

1,213 lb

Typical weight of an F1 car—less than half the weight of a Mini

320°F

Temperature that tires can reach

212°F

Optimal operating temperature for dry-weather racing tire

124 mi

Maximum life expectancy of an F1 car tire

220

Number of different materials used in a tire

0.725 psi

Degree of pressure change that can ruin performance of an F1 tire

81 mph

Speed an F1 car needs to achieve to produce downforce equal to weight of the car

3

Multiples of car's weight that an F1 car can produce in downforce at high speed

199

Juicing the Ball:
Baseball, Tennis and Velocity

115 mph

Approximate average speed of men's 1st serve at Wimbledon

156 mph

Current record for the fastest men's serve

The speed of a ball is vitally important. Faster balls travel farther, cross the same distance in less time, reduce the time available for an opponent to move into position and may even outpace the reaction time achievable by an opponent. The speed of a ball is influenced by many factors, from air humidity and altitude (see page 108) to the coefficient of restitution (the bounciness of the ball). Looking at baseball and tennis, it is possible to see the difference that differing applications of ball speed can make.

In tennis, a faster ball leaves less time for your opponent to react. The fastest balls are produced in the serve; the faster you can hit your serve, the further back your opponent has to stand to have any chance of reacting to and reaching the serve, and thus the bigger an advantage you have in coping with the return. In fact as the figures to the right reveal, the difference between an average serve and the fastest serve in the professional game can make the difference between reacting to a serve in good time and missing it altogether.

0.343 s

Approximate time for 155 mph serve to reach other side of court

0.462 s

Approximate time for 115 mph serve to reach other side of 3.77 m court

139

Home runs scored by National League hitters in 1919

460

Home runs scored by National League hitters in 1921

1919 **139 : 460** 1921

In baseball, a faster ball is harder to hit, as we saw on page 20. But a ball that travels faster will also go farther, making it easier to hit a home run. One way to make a faster ball is to make it bouncier by subtly altering its composition and construction. This is known as "juicing the ball," and many baseball historians believe that the major league authorities did just this in the 1920s to boost scores, making the game more attractive to pull in punters disillusioned by the Black Sox gambling scandal. Certainly the stats of home runs scored between 1915 and 1930 seem to back this up.

1,565

Total home runs scored across the league in 1930

384

Total number of home runs scored across the league in 1915

3.5

Additional runs per game scored by teams in 1930 compared to 1915

Formula 1

2012 FORMULA 1 JAPANESE GRAND PRIX

CIRCUIT

Suzuka is a high-speed circuit with a top speed of 315 km/h (196 mph) and an average speed of about 230 km/h (143 mph). The circuit length is 5.81 km (3.61 mi) and it has a varied mix of high- and low-speed corners. A total of 18 corners, comprised of 10 right turns and 8 left turns

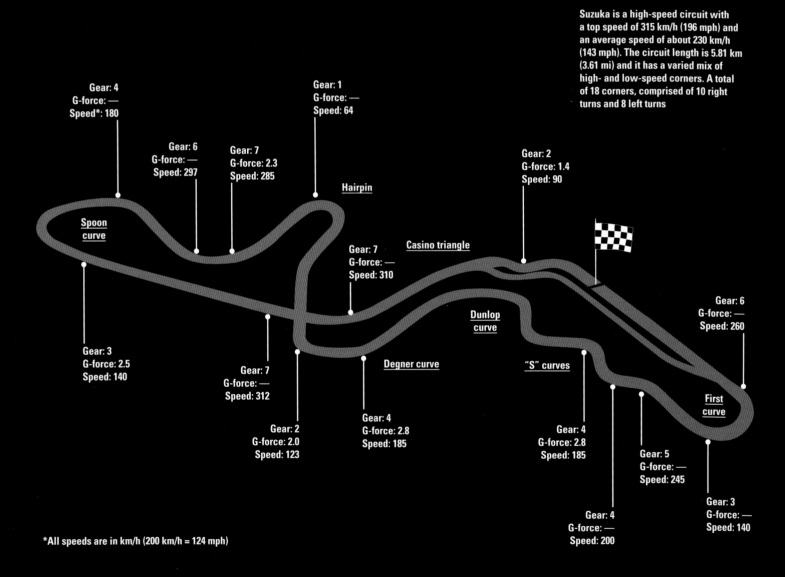

Gear: 4
G-force: —
Speed*: 180

Gear: 1
G-force: —
Speed: 64

Gear: 6
G-force: —
Speed: 297

Gear: 7
G-force: 2.3
Speed: 285

Hairpin

Gear: 2
G-force: 1.4
Speed: 90

Spoon curve

Gear: 7
G-force: —
Speed: 310

Casino triangle

Gear: 3
G-force: 2.5
Speed: 140

Dunlop curve

Gear: 6
G-force: —
Speed: 260

Gear: 7
G-force: —
Speed: 312

Degner curve

"S" curves

First curve

Gear: 2
G-force: 2.0
Speed: 123

Gear: 4
G-force: 2.8
Speed: 185

Gear: 4
G-force: 2.8
Speed: 185

Gear: 5
G-force: —
Speed: 245

Gear: 4
G-force: —
Speed: 200

Gear: 3
G-force: —
Speed: 140

*All speeds are in km/h (200 km/h = 124 mph)

CAR

The car must be no more than 180 cm (71 in) wide. Bodywork ahead of the rear wheel centerline must be a maximum of 140 cm (55 in) wide. Bodywork behind it must be no more than 100 cm (39 in) wide. Front and rear overhangs are limited to 120 cm (47 in) and 60 cm (24 in) respectively from the wheel centerlines. A typical car will be about 4,635 mm (182 in) long, 1,800 mm (71 in) wide and 950 mm (37 in) high. Cars must weigh at least 640 kg (1,410 lb) (including the driver) at all times

ENGINE

Formula 1 engines may be no more than 2.4 l (146 ci) in displacement. They must have eight cylinders in a 90-degree formation, with two inlet and two exhaust valves per cylinder. They must be normally aspirated, weigh at least 95 kg (209 lb) and be rev-limited to 18,000 rpm

BRAKES

Each wheel must have no more than one brake disc of 278 mm (10.9 in) maximum diameter and 28 mm (1.1 in) maximum thickness. Each disc must have only one aluminium caliper, with a maximum of six circular pistons, and no more than two brake pads

COCKPIT

Once strapped into the car with all his safety gear on, the driver must be able to remove the steering wheel and get out within 5 seconds, and then replace the steering within a further 5 seconds. The car's survival cell structure, designed to protect the driver in the event of an accident, must extend at least 300 mm (11.8 in) beyond the driver's feet, which must not be forward of the front-wheel centerline

TIRES

Formula 1 cars must have four uncovered wheels, all made of the same metallic material, which must be one of two magnesium alloys specified by the FIA. Front wheels must be between 305 mm (12 in) and 355 mm (14 in) wide, the rears between 365 mm (14.4 in) and 380 mm (15 in). With tires mounted, the wheels must be no more than 660 mm (26 in) in diameter (670 mm/26.4 in with wet-weather tires). Measurements are taken with tires inflated to 1.4 bar (20.3 psi). Tires may only be inflated with air or nitrogen

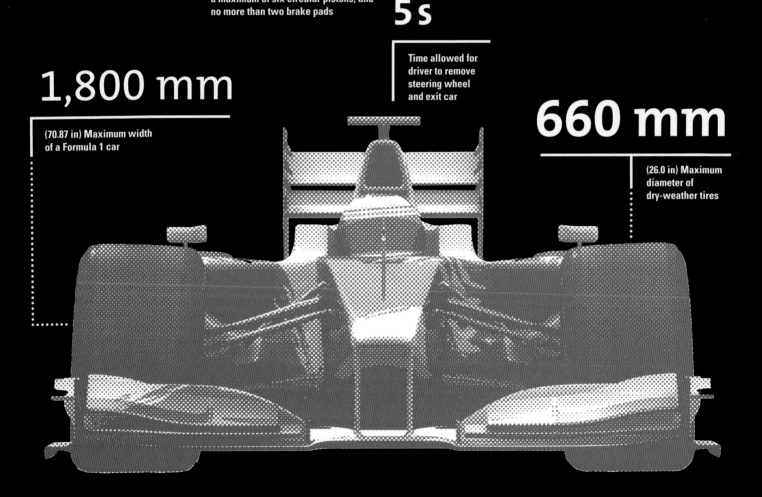

1,800 mm

(70.87 in) Maximum width of a Formula 1 car

5 s

Time allowed for driver to remove steering wheel and exit car

660 mm

(26.0 in) Maximum diameter of dry-weather tires

VO$_2$ the Max:
The Respiratory System

All sports come down to a few basic principles of biology. In order to expend physical effort, the body must burn energy-providing molecules called Adenosine Triphosphate (ATP). Each cell in your body has around a billion ATP molecules, which are used up and replaced every 2 minutes. Every day you produce and use up a volume of ATP equivalent to half your body weight.

There are two ways to burn ATP—with oxygen (known as aerobic respiration) and without (anaerobic respiration). Aerobic respiration produces water and carbon dioxide as waste products; the carbon dioxide is picked up by the blood and taken to the lungs where it is expelled when you breathe out. The body can cope with as much carbon dioxide as the muscles can produce. Once the body is unable to get sufficient oxygen to the muscle cells, however, they start to burn ATP anaerobically, which results in a waste product called lactic acid. This cannot be excreted as fast and quickly builds up in the muscles, causing the burning sensation familiar to anyone who has pushed themselves to the limit.

This means that the availability of oxygen is the ultimate limiting factor for exercise, which is why the best single measure of fitness and aerobic endurance is VO$_2$ max, or maximal oxygen uptake. It is often expressed in terms of volume of oxygen available for use per unit of body weight per minute, or ml/kg/min. Measuring it is a complicated business, usually involving a mask and a treadmill, with the test subject being required to work up to the highest level of exercise they can sustain and carrying on until their anaerobic limit is reached, which can be quite painful.

Very high VO$_2$ max levels are what set elite endurance athletes apart from the rest of us. Your personal level is mainly determined by your genes, but VO$_2$ max can be increased through fitness training, and is generally much higher in men than women, and in young people compared to older people.

40

Average number of breaths/min taken in heavy exercise

10–20

Number of breaths/min taken by a young adult male at rest

7 L
(7.4 qt) Maximum volume of air lungs of average adult male can hold

3.3 L
(3.5 qt) Typical volume of air exchanged in each breath

2 L
(2.1 qt) Minimum volume of air remaining in lungs after breathing out as hard as possible

c. 120 L/min
(c. 127 qt/min) Air circulation rate during heavy exercise

81 m²
(872 ft²) Surface area of lungs, equivalent to ⅓: area of a tennis court

MAX VALUES VO₂ (ML/KG/MIN)

	35–40	**Average untrained healthy male**
	27–31	**Average untrained healthy female**
	85	**Elite male runner**
	88	**Five time Tour de France winner Miguel Indurain**
	97.5	**Norwegian cyclist Oskar Svendsen (highest ever measured)**
	180	**Thoroughbred horse**
	240	**Siberian huskies running in the Iditarod Trail Sled Dog Race**

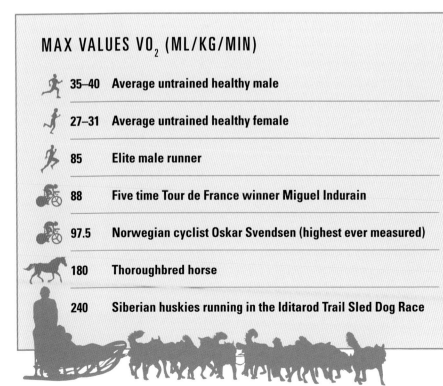

Blood and Guts: The Circulatory System

3 Supertankers that could be filled by your heart in the course of a lifetime

Apart from the respiratory system (the lungs and airways, discussed on the previous pages), the other body system that rules athletic ability is the cardiovascular system: the heart and blood vessels. Your heart is a prodigious organ capable of feats far beyond the capabilities of most modern-day technology—if you want to know how hard it works, try using a teacup to empty a bathtub full of water in 15 min. Now try repeating that every quarter of an hour for 80-odd years. But an athlete's heart can be up to three times bigger than an average person's.

Perhaps even more remarkably, the heart has an amazing ability to grow and shrink with exercise. A professional cyclist's heart almost halves in size when he stops training. These significant changes can happen with startling rapidity. A study of swimmers and cross-country runners at St. Louis University found that the swimmers increased the muscle mass of a key part of the heart by 23% after one week of hard training, while the runners

lost 38% of the mass of the same heart muscle just three weeks after stopping training.

A bigger heart is precisely the target for athletes undertaking aerobic training regimes. The biomechanics of heart pumping mean it is more efficient to pump a greater volume of blood with each stroke than to keep the stroke volume steady and use more strokes—in other words, pumping more is better than pumping faster. The maximum heart rate of an elite endurance runner is similar to that of a couch potato, but massively more blood is pumped with each stroke. A corollary of this is that the resting heart rate of an athlete is quite low, because fewer strokes are needed to meet resting requirements. An unfortunate and sometimes tragic side effect of this enlargement of the heart (a healthy condition known as "athlete's heart") is that it can exacerbate and trigger underlying heart problems, leading to the sudden collapse and death of what are otherwise extremely fit young men and women.

2.5

Heart beats in the average lifetime, in billions

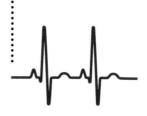

72 bpm

Average resting heart rate for a healthy adult

c. 30 bpm

Resting heart rate of elite marathon runner

40–50 bpm

Average resting heart rate for an athlete

EVERYDAY AMAZING FEATS OF THE AVERAGE PERSON'S HEART

- **18 days:** time it would take your heart to fill an Olympic-sized swimming pool
- **1 million barrels:** amount of blood pumped in a lifetime
- **100,000:** heartbeats per day
- **35 million:** heartbeats per year
- **2.5 billion:** heartbeats per lifetime
- **32 km (20 mi):** distance a truck could drive with the energy the heart expends each day—enough to drive to the Moon and back over a lifetime

46%

Average increase in mass of left ventricle (the part of the heart that does the most work and is most important to cardiovascular fitness) in athletes compared with non-conditioned control subjects

33%

Average increase in the volume of the left ventricle, with a complementary increase in stroke volume

The Speed of Thought:
Fencing, Tennis and Reaction Speeds

250 mph

Conduction speed
of nerve impulses

With balls and shuttlecocks traveling at around 200 mph, sports pushes the limits of human neurophysiology. The ultimate limiting factor in an athlete's ability to cope with high-speed play are the physiological speed limits of his or her nervous system. Reacting to an incoming tennis serve or the lunge of a fencer requires a series of sequential actions and processes to occur in the nervous system, and although nervous impulses move as fast as lightning, they do nonetheless take time.

A nerve cell or neuron looks a bit like a starfish, but with one arm much longer than the rest. Nerve signals are electrical impulses created by the rapid movement of charged ions across the membranes of the cell; they ripple up and down the arms of the neuron as very fast moving waves, which travel faster than Formula 1 racing cars. Where two neurons meet is a junction called a synapse, and the two cells are separated by a synaptic gap. When a nerve signal arrives at the synaptic junction it has to jump across, which is accomplished by chemical signaling.

The simplest physical responses, like the knee reflex you can see if you tap just below the kneecap, involve a simple loop of neurons running from the knee to the spine and back again. But more complex activities, such as parrying a lunge or hitting a backhand slice return of serve, require the nerve signal to travel all the way to the brain, where it must be processed, a response (i.e., a decision on what to do) generated and the response sent to the cerebellum. Cerebellum means "little brain"—it is the mass of tissue at the base of the brain that controls movement sequences, especially learned ones such as tennis strokes or fencing moves. The cerebellum implements a mental program, dispatching nerve signals to the relevant muscles, and these signals have to travel back down the nerve fibers to their targets before movement finally takes place.

Each step in this chain of events takes just milliseconds, but they all add up. It takes around 0.1–0.2 s to perceive a stimulus, and another 0.1–0.3 s to react to it.

0.043 s

Time for information about the velocity and trajectory of a baseball to be sent from the retina to the higher visual cortex, the portion of the brain dealing with visual information

0.0005 s

Synaptic delay—time for neurochemical transmitters to cross synaptic gap

356.49 m/s

Mean reaction time of Spanish national men's fencing team

399 m/s

Mean reaction time of Spanish national men's karate team

The Big Blue: Free Diving and the Diving Reflex

70°F

Temperature limit below which the diving reflex is triggered

One of the more extreme sports on the planet is free diving. Essentially free diving is just swimming under water while holding your breath, a state technically known as apnea. There are many different classifications of apnea, however, with different records for each one. All of them depend on a physiological reflex common to all mammals, known as the Mammalian Diving Reflex (MDF).

The MDF is triggered by immersing your face/body in water below 70°F—the colder the water, the more powerful the reflex. The body responds by automatically slowing the heart rate, redirecting blood from less important organs to the brain and other important ones, releasing booster blood stores from the spleen and triggering other responses that prolong the length of time you can hold your breath. Marine mammals with highly evolved versions of the MDF can achieve almost unimaginable dive depths and times.

Taking advantage of the MDF, humans can also surpass their apparent limits. If you try holding your breath now, while reading this, you may struggle to get past 60 seconds. If you dive under water you will probably be able to manage 2 min, thanks to the MDF. Under water apnea specialists make use of MDF to compete in one of the most dangerous sports in the world. The safest form is probably static apnea, which is basically floating motionless face down in the water. Then there is dynamic apnea, with and without fins, where swimmers compete to see how far they can swim horizontally while holding their breath. The really dangerous categories are the depth ones, where divers compete with or without weighted sleds, known as variable and constant weight apnea respectively, to see how far down they can get. The most dangerous of all is no limits apnea, where the only rule is that you are not allowed to take any external air source down with you.

90%

Percentage by which heart rate is slowed in some marine mammals due to MDF

10%–20%

Percentage by which heart rate in untrained humans is slowed as a result of diving reflex.

2 min

Average time someone swimming under water can hold their breath

50%

Diving-reflex-related slowing of heart rate in trained individuals

22 min 22 s

World record for breath-holding by a human, achieved by Tom Sietas in 2012

6,234 ft

Record depth for Cuvier's beaked whale, deepest dive confirmed of any air-breathing animal

702 ft

World record depth achieved by a human without external air source, by Herbert Nitsch, the "world's deepest man," in a no-limits apnea dive at Spetses in Greece in 2007

All Things Ice: Speed Skating and Ice

16°F

Temperature
of ice for hockey

22°F

Temperature of ice
for figure skating

Humans have contrived to turn almost every environment on Earth into a sporting arena in one form or another, whether challenging one another to dive deepest in caves, jump from the highest balloon or walk across Antarctica in winter. A good example is the collective national mania that descends upon the Netherlands every time they have a particularly brutal winter. As the temperature drops, the Dutch hope that it will get cold enough to freeze the Frisian waterways so that they can hold a unique ice marathon called the *Elfstedentocht*, the "Eleven Cities Tour." Only if the ice is declared to be thick enough across the entire 200 km (124 mi) course can the event be staged, so it has only happened 15 times since its official inception in 1909, and not since 1997.

The course has been skated at up to 30 km/h (19 mph), speedier than Usain Bolt can run at his fastest, while short-course speed skaters can reach over 50 km/h (31 mph). Even ice hockey skaters routinely reach speeds faster than Bolt can sprint. Skaters achieve the highest speeds it is possible to reach with the human feet unaided, thanks to the unique properties of ice. Ice has very low surface friction, because the top few layers of water molecules vibrate so fast that although locked in a solid crystal matrix they exist in a kind of semi-liquid state. Accordingly, the forward momentum generated by a skater can be conserved by gliding on one skate while the other is lifted and reset to provide the next boost of thrust. This cannot be done on the track because of the much higher friction.

The bad news for the Dutch and other fans of the Elfstedentocht is another set of numbers—those coming out of the the Intergovernmental Panel on Climate Change (IPCC). The evidence shows that the Earth is warming due to manmade carbon emissions, and the IPCC predicts temperature rises of 1.1°–6.4°C (2°–11°F) by 2100. The graph on the next page shows how the odds of hosting an Elfstedentocht are likely to decrease according to the upper, median and lower ends of the IPCC predictions.

20 mph

Top speed of
an NHL player

34.03 s

World record for 500 m
(1,640 ft) speed skating,
set by Canadian Jeremy
Wotherspoon in 2007

Average speed of
Wotherspoon—the
fastest a human can
travel using only his feet

33 mph

6 hr 47 min

Fastest-ever time, by Dutch
professional skater Evert van
Benthem in 1985

200 km

(124 mi) Length of the Elfstedentocht,
national sporting event in the Netherlands

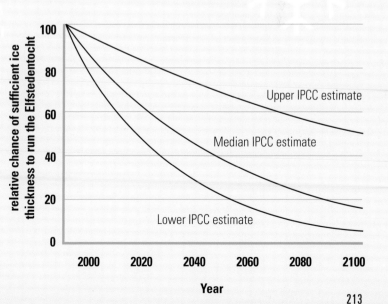

relative chance of sufficient ice
thickness to run the Elfstedentocht

100

80

60

40

20

0

Upper IPCC estimate

Median IPCC estimate

Lower IPCC estimate

2000 2020 2040 2060 2080 2100

Year

Body Heat:
Temperature and Performance

-256°F

Another limiting factor for athletes is temperature. Athletes compete across almost the entire range of temperatures the Earth has to offer. Cold-water swimmers, cross-country skiers and dogsled racers compete in the frozen north, while ultra-marathon runners punish themselves in the heat of African and American deserts. Surprisingly, it is not the environmental—or ambient—temperature that really matters, but the internal body temperature.

As mammals, humans have a highly evolved and highly effective system of thermoregulation—the process that keeps body temperature within sustainable limits. This is fortunate, because the internal temperature range at which the body can survive, let alone perform effectively, is relatively narrow. Below 95°F or above 107°F and you risk collapse and death, yet during exercise—running, for instance—body temperature quickly rises to 102°F. Even in hot conditions most runners are able to maintain their temperature at this level, thanks to sweating, but in particularly humid conditions body temperature may climb to 104°F, which is a kind of physiological cut-off point. At this temperature brain function is affected, leading to psychological and neurological deficits, such as partial paralysis, confusion and lack of co-ordination. These usually force runners to stop exercising, allowing their body temperature to drop, and in fact most runners will moderate their pace before reaching this limit.

At the other end of the scale, extreme cold triggers a survival mechanism called cold shock. This is associated with exposure to cold water rather than cold air, because water is a far better conductor of heat than air. Cold-water swimmers overcome it by habituating themselves to the cold, but for people who fall into freezing water it is the cold shock response that may inadvertently lead to death. Even in freezing water, loss of heat by conduction takes at least 30 min to result in a life-threatening drop in body temperature. But immersion in freezing water can cause hyperventilation and prevent effective swimming, which increase the likelihood of drowning. Sudden immersion in cold water also greatly increases the risk of heart attack.

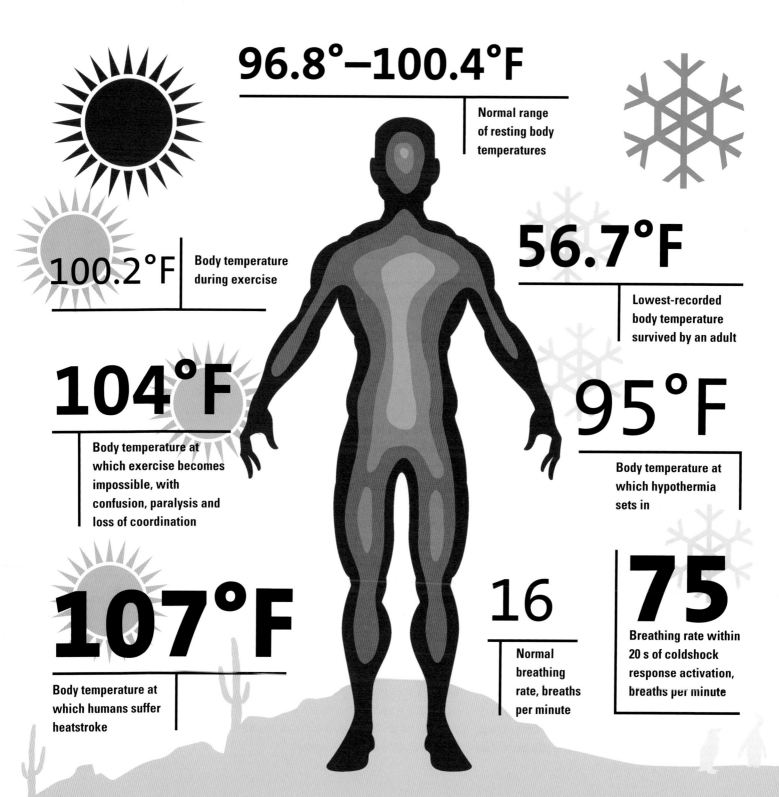

96.8°–100.4°F
Normal range of resting body temperatures

100.2°F Body temperature during exercise

104°F
Body temperature at which exercise becomes impossible, with confusion, paralysis and loss of coordination

107°F
Body temperature at which humans suffer heatstroke

56.7°F
Lowest-recorded body temperature survived by an adult

95°F
Body temperature at which hypothermia sets in

16
Normal breathing rate, breaths per minute

75
Breathing rate within 20 s of coldshock response activation, breaths per minute

Dry as a Bone: Hydration and Performance

1896

Year when water was not made available to marathon runners during the race

Water loss through sweating and exhaling is the primary mechanism by which the body regulates temperature. Human beings cannot survive in the unusual environment of the Naica Cave in Mexico, thanks to its unique combination of extremely high temperature and extremely high humidity. The measure that combines the two factors is known as wet bulb temperature, and the Naica Cave is the only place yet found on Earth where the wet bulb temperature is above the fatal 95°F limit. At this temperature the surface of the lungs becomes the coolest part of the human body and it is no longer possible to shed excess body heat through exhaling water vapor, leading to deadly heat buildup and hyperthermia.

Fortunately everywhere else on the planet is below this wet-bulb limit, including the places where athletes run endurance races. In these places athletes can shed heat through sweating and breathing and are able to maintain a core temperature no higher than 102.2°F.

Doing so, however, involves high rates of water loss, and athletes must rehydrate to be able to compete. According to traditional theories of sports physiology, athletes don't just lose water but electrolytes as well. Electrolytes are the dissolved salts that the body's nerves and muscles use to activate muscle contractions, and a multi-billion dollar industry has built up on the basis that exercising causes you to lose significant amounts of electrolytes and that electrolyte-rich sports drinks boost performance levels by replacing these.

In fact there is little hard evidence to back this up. The quantity of electrolytes lost in sweat is minuscule, especially compared to the amount of water lost. Heavy sweating is more likely to increase the relative concentration of electrolytes in your body. What is more, only specific muscles seem to be affected by cramps (whereas a systemic shortage of electrolytes should affect all muscles). Cramps are almost certainly caused by fatigue, not electrolyte depletion.

1936

Year when athletes were offered drinks every 3 km (1.9 mi)

1.9 mi

60%–65%

Proportion of the adult male body that is water

50%–60%

Proportion of the adult female body that is water

Weight lost by NHL players over the course of an average game, mostly due to water loss

5–8 lb

Factor by which it may be necessary to increase water consumption to survive in desert

2%–5%

Degree of dehydration of elite endurance runners

1 qt/hr

Rate at which the body can lose water in extreme desert conditions —equivalent to sweating a soda can's worth of liquid every 20 mins

Sky High: Skydiving and Terminal Velocity

▲ **2.5 hours**

Time taken to reach 24 mi altitude

▼ **335 s**

Time taken to return to Earth from the height of 24 mi

The precise opposite of free diving is skydiving. Here the challenge is not how deep you can go, but how high, although at extreme altitudes some of the risks are the same. It is not possible to breathe in the hostile environment and the change in pressure can cause fatal decompression events.

The thrill for many sky divers is the feeling of weightlessness in freefall, but although they may feel the buffeting of wind resistance, they only experience the sensation of acceleration, and therefore speed, for the first 7 seconds or so. This is because a falling body quickly reaches terminal velocity: the speed at which the upward force from aerodynamic drag and wind resistance equals the downward force of gravity. The precise magnitude of this speed depends on a number of factors: the mass of the body, the cross-sectional surface area it presents in the direction of travel (i.e., spreading out arms and legs and wearing flappy wingsuits increases surface area and drag and reduces the terminal velocity) and the density of the air through which it is falling.

Even sky divers adopting a streamlined "bullet posture" with their heads pointing down and their arms tucked into their sides cannot top 200 mph when jumping from normal altitudes within the Earth's troposphere, but if you go high enough the air gets thin enough for you to accelerate to tremendous speeds. This was the reasoning behind the world record altitude sky dive of Felix Baumgartner in 2012, during which he broke the sound barrier before slowing down once he reached the thicker parts of the atmosphere.

Lighter bodies with a higher ratio of surface area to mass will reach terminal velocity sooner and slower than heavier ones. This is part of the reason that cats can survive falling from skyscrapers but humans cannot. Within five stories a cat reaches terminal velocity. No longer feeling acceleration, the cat relaxes and adopts a spread-eagle stance that spreads its impact force over a wider area.

834 mph

Maximum velocity Felix Baumgartner reached (faster than the speed of sound in the thin atmosphere)

40 s

Time taken to reach this speed

120 mph

Terminal velocity for sky diver falling with arms and legs spread

321 mph

Terminal velocity for a sky diver in streamlined posture below c. 1.9 mi altitude

5 stories

Height from which a falling cat reaches terminal velocity

55 stories

Height from which a person would need to fall to reach terminal velocity (548 ft)

`07.00`

7

Number of seconds for a falling body to reach 95% of terminal velocity

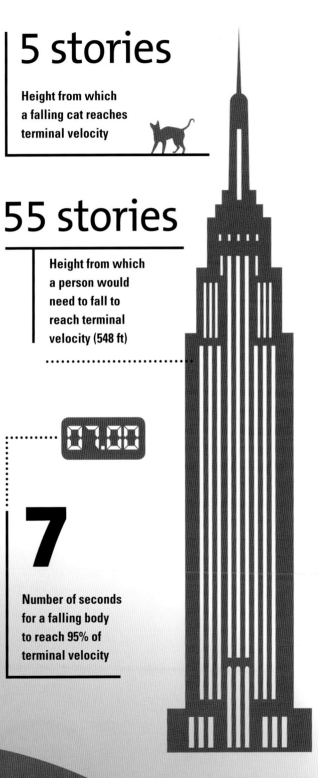

219

The Outer Limits: Physiology, Biomechanics and Sports

1970s

Era when horse and dog races reached their limit of performance

Many sports fans subscribe to the fallacy that since records have been broken in the past, and are still being broken, they will inexorably tumble in the future. For instance, it is assumed that Usain Bolt's exciting achievements in successively beating his own world records show that the time for the 100 m will continue to come down: if Bolt doesn't keep breaking the record, the next generation of sprinters will be faster.

While there may well be room for improvement in many, perhaps all, world records, there will also be limits, set by human physiology and biomechanics. Assuming that these are not augmented, either pharmacologically or by engineering (genetic or cybernetic), is it possible to predict how much will be shaved off world records in the future?

It turns out that it may be possible, thanks to mathematics. By looking at trends in the progression of records, analysts at the French Institute of Sports were able to assess how close to the limits of possible performance the athletes of the past have been, and how close they may come in the future. Using similar methods, a 2008 study by Mark Denny of Stanford University, published in the *Journal of Experimental Biology*, compared human running performance to the racing times of horses. Denny found that horses seem to have already reached the limits of their performance, and was able to use the curve generated by equine records as a model to predict the limits of human performance.

Years that the record for the Kentucky Derby has stood

>30

9.48 s

Peak possible time for 100 m predicted by Mark Denny

`09.48`

Improvement expected from each successive generation

0.07–0.10 s

2030

Probable date by which limit for the 100 m will be reached, according to Denny

3,260

Number of world records set between first modern Olympics in 1896 and 2008

2027

Year by which half of events will have reached performance ceiling, according to study by French Institute of Sport

2060

Year by which all athletics world records will have hit a ceiling, according to study by French Institute of Sport

Index